Edwin Moses Hale

A systematic Treatise on Abortion

Edwin Moses Hale

A systematic Treatise on Abortion

ISBN/EAN: 9783337034047

Printed in Europe, USA, Canada, Australia, Japan

Cover: Foto ©berggeist007 / pixelio.de

More available books at **www.hansebooks.com**

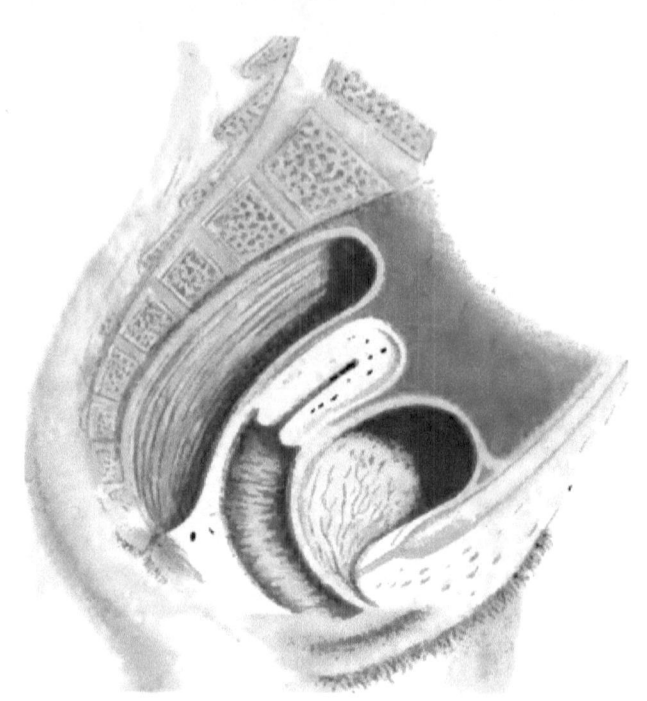

A

SYSTEMATIC TREATISE

ON

ABORTION:

BY

EDWIN M. HALE, M.D.,

PROFESSOR OF MATERIA MEDICA AND THERAPEUTICS IN HAHNEMANN MEDICAL COLLEGE;
ETC., ETC.

CHICAGO.
C. S. HALSEY, 147 CLARK STREET.
1866.

Entered according to Act of Congress, in the year 1866,

BY C. S. HALSEY,

In the Clerk's Office of the District Court, for the Northern District of Illinois.

CHURCH, GOODMAN AND DONNELLEY, PRINTERS, 51 and 53 LaSalle Street, Chicago.

THIS
VOLUME
IS
INSCRIBED
TO
A. E. SMALL, A.M., M.D.,
OF CHICAGO:

A MOST EXCELLENT PHYSICIAN,
ONE OF THE BEST OF MEN,
AND TRUEST OF FRIENDS.

PREFACE.

In the Eighth Volume of the *North American Journal of Homœopathy* (1860), an article appeared bearing the title, "Abortion: its Prevention and Treatment," in which the writer ventured to call the attention of his professional colleagues to medicines not heretofore used by members of the Homœopathic school, and which he believed to be of considerable value as remedial agents for the removal of certain abnormal conditions of the organs of generation, which are likely to cause an arrest of development, or premature expulsion of the product of conception. The interest that this article aroused in the professional mind was such that the writer felt justified in enlarging and amending the original paper. It was then issued by the present publisher, in a pamphlet of twenty-two pages, entitled "*The Homœopathic Treatment of Abortion*," etc.

This small *brochure* was prefixed by a Prefatory Letter from Dr. R. Ludlam, in which he kindly commended it to the notice of Homœopathic physicians.

That edition of the monograph having been exhausted, the writer engaged in a more thorough and systematic study of the subject. Five years having elapsed since the appearance of the original paper, he had gained increased practical experience upon the subject. He has, therefore, attempted to embody these results in the Treatise herewith presented to the general medical profession, and especially to physicians of the Homœopathic school.

He sincerely hopes this volume will form a useful addition to our medical literature.

CHICAGO, *April* 2, 1866.

CONTENTS.

	PAGE
INTRODUCTION	xiii

PART I.—STATISTICS OF ABORTION.

Of Foreign Countries	19
Of New York City	20
Of Boston	22
Of Chicago	23
As to the period at which Abortion occurs	25
Of Criminal Abortion	27

PART II.—ETIOLOGY OF ABORTION.

TABULAR VIEW OF THE CAUSES OF ABORTION	33

SEC. I.—PREDISPOSING CAUSES.

Plethora	35
Anæmia or Chlorosis	35
Scrofula	36
Return of Menstrual Crisis	36

Zymotic Diseases:

Syphilis	37
Mercurialization	38
Variola	38
Asiatic Cholera	39
Yellow Fever	39

SEC. II.—LOCAL CAUSES.

Abnormal Condition of the Ovum and its Appendages	39
Moles	40
Hydatids	41
Fatty deterioration of the Chorion and Placenta	42
Congestion of the Placenta	43
Inflammation of the "	43
Placenta Previa	44

SEC. III.—CENTRIC CAUSES.

Emotional	45
Physical	45
Medicinal	46

viii.   CONTENTS.

	PAGE
Galvanism	48
Cerebro-Spinal-Meningitis	49

Sec. IV.—Concentric Causes.

Parotidean	55
Thyroidean	56
Mammary	56
Gastric	56
Dental	57
Renal	57
Vesical	57
Rectal	57
Vaginal	58
Ovarian	58
Uterine	59

Sec. V.—Functional Diseases of the Uterus.

Congestion	61
Leucorrhœa	62
" Cervical or Mucous	64
" " Sequelæ	65
" Vaginal or Epithelial	67
Gonorrhœa	69

Sec. VI.—Organic Diseases of the Uterus and Cervix.

Ulceration of the Cervix	71
Simple Granulating Ulcer	73
Varicose "	76
Fissured "	77
Follicular "	79
Phagœdenic "	80
Corroding "	80
Cancerous "	80
Syphilitic "	80

Sec. VII.

Induration of the Cervix Uteri	82
Displacements of the Uterus	84
Anteversion	84
Prolapsus	84
Retroversion	84
Death of Embryo	87
Coitus	87
Instrumental Irritation	88
Ovarian	89
" Irritation	89
" Congestion	90
" Inflammation	90

Sec. VIII.

Medicinal Causes 90

PART III.—GENERATION:
Symptoms, Diagnosis, Pathology, Mechanism and Prognosis of Abortion.

Sec. I.—Generation.

To date of Placental Attachments 115
" Viability 117
" Delivery 118
Dimensions and Weight of the Fœtus at the different periods of Uterine Life 119

Sec. II.—Symptoms of Abortion.

Premonitory 123
Actual 124
    Chills 124
    Pain 125
    Hæmorrhage 127
    Fever 130
    Complications 131
    Discharges 131
Subsequent 131

Sec. III.—Diagnosis of Abortion.

Metritis 134
Peritonitis 135
Dysmenorrhœa 135
Dysentery 135
Congestion of Uterus 136
Hydrorrhœa 137
Hæmorrhage 138
Retention in Utero of the Ovum and Appendages 140

Sec. IV.—Pathology and Mechanism of Abortion.

Process of Abortion 142
After Conception 143
" twenty days to the third month 143
At sixth month 146

Sec. V.—Prognosis of Abortion.

When favorable 148
" unfavorable 149
Natural abortions 150
Immediate consequences 151
Remote " 151

PART IV.—TREATMENT OF ABORTION.

Sec. I.—Preventive Treatment.

	PAGE
Of Plethora	155
Anæmia or Chlorosis	159
Scrofula	162
Return of Menstrual Crisis	162

Of Zymotic Diseases:

Syphilis	163
Mercurialization	164
Variola	164
Scarlatina	166
Diphtheria	168
Cholera	170

Sec. II.—Of Local or Organic Diseases.

Malformation of the Ovum	171
" " Membranes	171
Placenta Previa	172
Organic Disease of Placenta	173
Fatty Degeneration	173
Hydatid "	173
Calcareous "	175
Molar "	175

Sec. III.—Of Reflex Causes.

Centric	176
Emotion, Fright, etc	176
Blows, Injuries, etc	177
Medicinal	178
Concentric	178
Parotidean Irritation	178
Thyroidal	179
Mammary	179
Dental	180
Gastric	182
Rectal	183
Vesical	186
Vaginal	187
Hysterical	189
Epilepsy	189
Falls, Jumping, Blows, etc	189

Sec. IV.—Of Functional Diseases of the Uterus.

Congestion	191
Inflammation	192
Leucorrhœa	192
Cervical	192

	PAGE
Vaginal	194
Gonorrhœa	197

Sec. V.—Of Ulceration of the Os and Cervix Uteri.

Simple Granulating Ulcer		204
Varicose	"	204
Fissured	"	205
Follicular	"	206
Phagædenic	"	206
Syphilitic	"	208
Resumé		209
Diphtheria		209
Ovarian Diseases		211

Sec. VI.—Of Uterine Displacements.

Prolapsus	213
Anteversion	214
Retroversion	214

Sec. VII.—Remedial Treatment.

Medicinal	217
Mechanical	224

Sec. VIII.

Conduct of the Physician	234

Sec. IX.

Examination of the Patient	237

Sec. X.

Management of Labor	240

Sec. XI.—Sequelæ of Abortion.

Post-partum Treatment	248
Postural	248
Dietetic	251
Medicinal	252
Of Pelvic Cellulitis	253
Hypertrophy of the Uterus	253
Fistula	253
Inflammation of the Uterus	253
Puerperal Metritis	254
" Peritonitis	254
Phlebitis	254
Dropsy	254
Paralysis	254
Mental Aberrations	254
Chronic Menorrhagia	255
Mechanical Treatment	256
Of Prolapsus	256
Retroversion	256

PART V.—OBSTETRIC ABORTION.

Sec. I.—Obstetric Abortion.

	PAGE
Necessity of Premature Labor	263
Methods employed—Stillette	271
Sponge-tent	271
Caoutchouc Bags	271
Tampon	272
Colpeurynteur	273
Water-douche	273
Ergot	275
Cupping-glasses	275
Galvanism	277

Sec. II.—Fœtal Abortion.

Methods employed	280
Flexible Bougie	281
Catheter	281

Sec. III.—Embryonic Abortion.

Methods employed	284
Flexible Catheter	284
Uterine Sound	287
Injections	288
Abortion **Forceps**	288

Sec. IV.—Ovular Abortion.

What it is	290
Utopian Theories	291
Duty of the Physician	292
Different methods	293
Theory of Impregnation	295
Character of Spermatozoa	299
Prophylaxis of Conception—agents employed	303
Tabular View of Uterine Contents and Synopsis of Treatment	309

PART VI.—JURISPRUDENCE OF ABORTION.

Sec. I.—Criminal Abortion.

A Lecture by A. E. Small, M.D.	313
" Charles Woodhouse, M.D.	320
Laws of the European Countries	321
" different States	323
General Index	339

INTRODUCTION.

The term ABORTION, is derived from the Latin word *aborto*, which means literally — *to bring forth before the time*. This broad definition allows of no restriction, aside from the one given in the literal rendering. The premature expulsion of the contents of the gravid uterus at any date prior to the end of the ninth month, or the normal expiration of pregnancy, is an abortion. In this work, the term has therefore been used in its broadest sense; but, as will be seen, I have divided the period of pregnancy into three natural divisions, based on the condition of the placenta and the fœtus. Four kinds of abortion are treated of, namely: (1) OVULAR, when the ovum is lost before it is impregnated. (2) EMBRYONIC, when the impregnated ovum is expelled before the placenta has formed its uterine attachment. (3) FŒTAL, when the expulsion occurs after the last date, and before the viability of the child; and (4) when the child is born capable of living, or viable, but before the end of a normal pregnancy.

This is not an arbitrary plan, because it is founded on natural changes and certain periods, which are well recognized in physiological science. Such an arrangement greatly facilitates a study of the subject in a methodical manner.

Hitherto no complete and systematic treatise on Abortion has appeared in the literature of any school of medicine.

The Allopathic branch of the profession, in which we would naturally look for a work on this subject, possesses but one volume, which treats of it in a monographic manner.

Dr. Whitehead's work on "Abortion and Sterility" was published in 1854. While it is a valuable work of reference on certain points, it is lacking in systematic completeness. It contains some suggestive statistics, and an **excellent elucidation of** many of the causes of abortion, with **their Allopathic treatment;** but beyond **this,** the author does **not extend the work.** No mention is made of the intermediate and remedial treatment of the **accident itself, or of its numerous** and important sequelæ.

The obstetrical works of Churchill, Ramsbotham, Tyler Smith, Simpson, **Cazeaux, Hodge, Meigs, Bennett,** Bedford, Gardner and others (Allopathic); King, Scudder, **and** Beach, (Eclectic); **and** Jahr, Leadam, **Pulte, Loomis, Small, Marcy,** Ludlam and **others** (Homœopathic); all contain **much in** relation to this subject.

I have drawn freely from **all these authorities,** selecting the practical and useful, and rejecting all that was irrelevant. Of all medical writers, Cazeaux seems to me to have treated the subject **of abortion in the most** systematic manner.

The various periodicals of our school have been **examined for clinical experience relating to** the treatment **of this accident.** Considerable practical information on this point has **been communicated** to me by my professional friends. **I am especially under** many obligations to Dr. R. Ludlam, Professor of **Obstetrics and Diseases of Women and Children in** Hahnemann Medical **College,** for his kind assistance **while in** the preparation of this work, **and** also for the use of his large obstetrical library, containing many rare and valuable volumes.

I am obliged to state, however, **that** on many points of importance **relating** to the *treatment* and *induction* of abortion I have had to rely almost wholly upon my own experience, and to ascertain **the most** practical **and** useful facts by careful experiment **and** patient investigation.

Abortion has grown to be a subject of such importan**ce to the** medical man, that its consideration should no longer **be** confined to the works devoted to obstetrics and diseases of women, wherein it **can but** be treated of in a manner not sufficiently complete to satisfy the student or investigating practitioner.

INTRODUCTION.

It must strike every observant and thinking physician that the day for large treatises, on such broad subjects as "Practice," "Materia Medica," "Surgery," "Obstetrics and Diseases of Women," has gone by. No medical writer can do justice to the range of subjects naturally included in any one of the above titles, if he is confined to one or even two volumes. Nothing short of an encyclopœdia of one of the above subjects would satisfy the requirements of the age, and as an encyclopœdia is made up of separate monographs, it will **greatly enhance the value of medical** literature **if a series of** exhaustive monographs appear which shall **do away with the necessity for such ponderous works.** Moreover, the accumulations of such valuable material relating to any one subject pertaining to practical medicine has become so extensive, and withal so scattered through the books, journals and other periodicals, that it is the duty of the medical writer to collect the material together, and put it in such a form as to be most available to the student or practitioner.

To this end the writer of this volume has directed his labors. He does not claim to present much original matter, although his observations and experience have enabled him to increase somewhat the common stock of information relative **to the etiology and** treatment of the accident.

He has drawn freely from all the standard medical works of the day, and from all sources which **seemed** to him reliable. Medical *facts* are common property, and **it is not** necessary to give further credit to authors and observers, **than is** given in the pages wherein mention is made of the sources of information.

This work is intended in no respect for public circulation, and the author would **be sorry** to think it should ever be perused by the prurient and immoral.

In order to render the work **as complete** as possible, it has been necessary to mention the various methods to be made use of for the induction of abortion for legitimate **purposes.** If this information shall be abused, and used for unlawful purposes, the blame must **rest where it** really belongs.

If I have not given minute indications for the use of each remedy mentioned in the following pages, it is because of the impossibility of so doing in a work of this kind, without swelling it to an inordinate size. The symptoms of, and special indications for, the use of each medicine can be found in the **several works on Materia Medica**, belonging to the Homœopathic school.

I believe that no physician in whose hands **this work may fall**, will consider the volume a superfluous addition to **our literature.**

PART I.

STATISTICS OF ABORTION.

STATISTICS OF ABORTION.

It will be appropriate, before we enter upon the consideration of the causes, pathology, treatment, etc., of Abortion, that we shall make ourselves acquainted with the statistics having a direct bearing upon the subject.

The first question which presents itself—*Is Abortion constantly increasing?*—is so important that we shall present the statistics in as complete a manner as is consistent with the plan of this work.

The following statistics are taken from an elaborate work* by Horatio B. Storer, M. D., of Boston. Writing of the frequency of abortion in our own country, he says:

"Statistics in this country are yet so imperfect that we are necessitated to a process of deduction. * * * If we find that in another country living births are steadily lessening in proportion to the population and to its increase—that natural and preventive causes are insufficient to account for this—while the proportion of still births and of known abortions is constantly increasing, and these last bear an evident yet increasing ratio to the still-births; that in this country the decrease of living births, and the increase of still-births, are in much greater ratio to the population, and the proportion of premature births is constantly increasing; and that these relations are constantly and yearly more marked, we are justified in supposing that abortions are

---
* Criminal Abortion in America.

at least as frequent with us, and probably more so. In many countries of Europe it has been ascertained that the 'fecundity' of the population, or the rate of its annual increase is rapidly diminishing.

"In Sweden it has lessened by a fifth; in Prussia, by a fourth; in Denmark and England, by a third, and in Russia, Spain, Germany and France, by a half, in *a single century!*

"In four departments of France, among which are two of the most thriving of Normandy, the deaths actually exceed the births!

"Again, as might have been expected, we find that the proportion of still-births, in which we must include abortions, as has hitherto been done, however improperly, in all extensive statistics, is enormous, and is steadily increasing; and while the proportion of still-births to the whole number is greatly increasing in Paris, as is the number of known abortions. * * At the Morgue, which represents but a very small fraction of the fœtal mortality of Paris, and in this matter almost entirely crime, there were deposited during the eighteen years preceding 1855, a total of 1115 fœtuses, of which 423 were at the full term, and 692 were less than nine months, and of these last, 519, or five-sixths, were not over six months, a large proportion of them showing decided marks of criminal abortion.

"We now turn to our own country, to which the city of New York holds much the same relation that Paris holds to France.

"Since 1805, when returns were first made to the Registry of New York, the number, proportionate as well as actual of fœtal deaths, has steadily and rapidly increased. With a population, at that time, (1805) of 76,770, the number of still and premature births was 47; in 1849, with a population at 450,000, the number had swelled to 1320."

In brief, while the ratio of fœtal deaths to the population was, in 1805, 1 to 1633.40, in 1849 it was 1 to 340.90; and when we consider that a large proportion of the reported premature births **must always be from criminal causes**; and that though almost all the still-births at the full time, even from infanticide, are necessarily registered, but a small proportion of the abortions and miscarriages occurring are ever reported to the proper authorities, it will be apparent that at the present moment the abortion statistics of New York are far above those of 1849. This the following table will show, as well as the fact that the ratio is steadily increasing:

	Total mortality.	Still-birth.	Ratio.
1804 to 1809	13,128	349	1 to 37.6
1809 to 1815	14,011	533	1 to 26.3
1815 to 1825	34,798	1,818	1 to 19.1
1825 to 1835	59,347	3,744	1 to 15.8
1835 to 1855	289,786	21,702	1 to 13.3
1856	21,658	1,942	1 to 11.1

"The frequency of abortions and premature births reported from the practice of physicians, and thus to a certain extent, but not entirely, likely to be of natural or accidental origin, is as follows:

"In 41,699 cases registered by Collins, Beatty, La Chapelle, Churchill, and others, there were 530 abortions and miscarriages. Here all the abortions were known: their proportion was 1 to 78.5.

"In New York, from 1854 to 1857, there were 48,323 births at the full time reported, and 1,196 premature. Here all the abortions were not known, probably but a very small fraction of them: the proportion was 1 in 40.4. In the seventeen years from 1838 to 1855 there were reported 17,237 still-births at the full time, and 2,710 still prematurely; the last bearing the proportion of 1 to 6.3. In the nine years from 1838 to 1847, there were 632 premature still-births, and 6,445 still at the full time: a yearly average of 1 in 10.2. In the eight years from 1848 to 1855, there were 2,078 premature still-births, and 10,792 still at the full time: an average of 1 in 5; while in 1856 there were 387 still prematurely and 1,556 at the full time, or 1 in 4.02.

"From these figures there can be drawn but one conclusion—that criminal abortion prevails to an enormous extent in New York, and that it is steadily and rapidly increasing. 'We cannot refer,' was well said by a former Inspector of that city,* 'such a hecatomb of human offspring to natural causes.'"

The same statistics also shew that the reported *early* abortions, of which the greater number of course escape registry, bear the ratio to the living births of 1 in 40, while elsewhere they are only 1 in 78.5: and finally, that early abortions, bearing the proportion to the stillbirths at the full time of 1 in 10.2 in 1846, had increased to 1 in 4.2 in 1856.

Almost doubling, therefore, as does New York, the worst of those fearful ratios of fœtal mortality existing in Europe, it is not strange that our metropolis has been held up, even by a Parisian, to the execration of the world : "On le voit (l'avortement)," says Tardieu, "en Amérique, dans une grande cité comme New York, constituer une industrie veritable et non poursuivie."

"In this description of New York," says Dr. Storer, "we have that of the country."

"In Boston, which for morals is supposed to compare favorably with any city of its size in the Union, undoubtedly more than a hundred still-births yearly escape being recorded; a large proportion of which, no doubt, results from criminal abortion."

In the State of Massachusetts it appears that during the fourteen years and eight months preceding 1855, there were recorded 4570 still-births, and 11,716 premature births and abortions, the ratio being one abortion to three still-births; or, in other words, it would appear from the statistics quoted, that the comparative frequency of abortions in Massachusetts is *thirteen* times

---

* Report of 1849.

as great as in the worst statistics of the city of New York!

It must not be forgotten that while nearly every still-birth at the full time is necessarily recorded, there must be but very few registrations of the premature births and abortions actually occurring. Few persons could have believed possible the existence of such frightful statistics. They should call the attention of the whole medical and legal world toward some plan to arrest such awful destruction of embryo human life.

In the great city of Chicago no registration of the births has ever been made; nor do the physicians make any returns relating to the still-births, miscarriages, etc. This is much to be regretted on many accounts. It does not do credit to the municipal authorities, to the influential citizens, nor to those physicians of the dominant school, who are supposed to have influence with those in power. It is alleged as an apology for this omission of a civil duty, that the State legislators have never passed a law of Registration of Births, etc.; but this is no valid excuse, when it is in the power of the city authorities to pass an ordinance which would answer every purpose of such a law.

To the above statistics of the *frequency* and *increase* of abortions, I will add those of Dr. Whitehead,[*] who asserts that "the number of pregnancies which each woman experiences, during the existence of her procreative aptitude, is about twelve, or one in every twenty months. This includes abortions, false conceptions, premature deliveries, and all having an unsuccessful issue, the average amount of which will be rather more than one and a half for each individual; or it may be

---

[*] On Abortion and Sterility, page 198.

stated, as a general rule, that *every seventh pregnancy* has a premature termination. These conclusions have been drawn from the subjoined facts.

"Two thousand married women in a state of pregnancy, admitted for treatment at the Manchester Lying-in Hospital, during parts of the years 1845 and 1846, were interrogated in rotation respecting their existing condition and previous history. Their average age at the time of inquiry was a small fraction below thirty years. The sum of their pregnancies already terminated was 8681, or 4.38 for each, of which rather less than one in seven had terminated abortively. But as abortion occurs somewhat more frequently during the latter than in the first half of the child-bearing period, the real average will consequently be rather more than one in seven. Of the individuals submitted to inquiry, 1253 had not then suffered abortion. The average age of these was 28.62 years; the sum of their pregnancies was 3906, or 3.11 for each person. The remaining 747 had already aborted once at least, and some oftener. Their average age was 32.08 years. The sum of their pregnancies was 5775, or 6.37; that of their abortions 1222, or 1.63 for each person. From the preceding statements it appears that more that thirty-seven out of every hundred mothers, experience abortion before they attain the age of thirty years."

Dr. Whitehead's observations do not accord with the popular idea, that early, especially first pregnancies, have more frequently a premature termination than those which come after. He is inclined to believe that the third and fourth, and subsequent pregnancies, and one or two of the last—namely, those which occur near the termination of the fruitful period—are most commonly unsuccessful.

At what period of pregnancy does abortion most frequently occur?

The following table, copied from Whitehead, throws some light on this question. It will be noted that the abortions here referred to, were those which were supposed to occur from diseased conditions of the mother or fœtus. Hereafter the subject of *criminal* abortions will be referred to. ABORTION (which term is here used in its most extensive signification) may occur at any period of utero-gestation. It will be seen that it is much more common at some stages of the process than at others, and is attended with different degrees of danger, according to the circumstances under which it occurs, the nature of the exciting cause being the most important. When it takes place before the end of the sixth month, it is invariably fatal to the offspring, either before birth, or in a short time after; and at any period before the completion of the process, it is more or less injurious to its well-being. Instances are on record, however, of children born during the early part of the seventh month, having lived in the enjoyment of tolerable health and constitutional vigor to a mature age.

" Abortion is at all times fraught with danger to the mother, and sometimes attended with fatal consequences. I give in the following table the respective periods of 602 cases of abortion, which have occurred under my own immediate observation. It may be noticed that each figure in the first column embraces a period of four weeks, extending from a fortnight before to the same length of time after the month indicated; and as abortions happening earlier than the seventh week of uterine life are so frequently and so nearly simulated, both in married and unmarried females, by certain uterine discharges, the result of disordered menstruation, events said to have taken place at this early period, except those wherein the escape of an ovum was undoubtedly found, have not been included in the report."

TABLE

Showing the Period of Pregnancy at which Abortion occurred in 602 cases, the relative number of Stillborn and Living Children, and the number living at the end of a month after birth.

Period of pregnancy at which abortion occurred.	Number of births at each period.	Number stillborn.	Number living at birth	Number living at end of a month
2 months	35	—	—	—
3 "	275	—	—	—
4 "	147	—	—	—
5 "	30	—	—	—
6 "	32	24	8	0
7 "	55	38	17	3
8 "	28	23	5	1
Total	602	85	30	4

"The fœtus of six months' growth is generally considered viable. Of the eight indicated in the table, as having been born alive, when born at this period, seven perished within six hours after birth, and one only attained to the age of ten days. Of the seventeen born alive at seven months, the majority lived over several days, and a few to the end of the third and fourth week. Three still survive, the youngest of whom is now aged nineteen months."

Whitehead does not mention the possibility or probability of an abortion occurring at the *first* month, or even at the *third week* of pregnancy. Yet it must be admitted by those who have investigated the causes of abortion, that it is by no means improbable, and very possible, for the abortions which occur at that early period to outnumber those of any other month.

If any disease of the uterus or contiguous organs, or any constitutional irritation sufficient to cause abortion exists, such causes are more likely to induce abortions at the FIRST *month than at any other period.*

Let us consider the matter logically. Intimately connected with the subject above treated of, is that of *criminal* abortion, and the period of pregnancy at which it generally occurs.

From Dr. Storer's work we learn that "It was Orfila's opinion that criminal abortion was most frequent in the first two months of pregnancy. This would naturally have been supposed to be the case, as then some doubt always might obtain regarding its existence, and the excuse that the measures resorted to were for the purpose of preventing ill effects from an abnormal menstrual suppression, would be more available. Devergie, on the other hand, was inclined to put the limits of greater frequency at from three months to four and a half; while Briand and Chandler thought the crime more common in the third month than the fifth, and in the last month much more frequent than even in the first or second. Tardieu also came to a similar conclusion. He ascertained that of 34 cases investigated by himself, 25 were in from the third to the sixth month, mostly in the third; 5 in the first two months; 4 in the second and eighth; or that the cases in the third month, or shortly after, were five times as numerous as at either an earlier or a later period, and nearly three times as numerous as in both combined. Upon examining the register of the Morgue, we find its statistics strikingly corroborative of this deduction. We have already seen that from 1837 to 1854 there had been deposited at the Morgue 692 fœtuses of less than nine months. Of these

23 were from the first to the second month;
79 " second " third "
108 " third " fourth "
158 " fourth " fifth "
150 " fifth " sixth "
97 " sixth " seventh "
48 " seventh" eighth "
29 " eighth " ninth "

"It has been stated that 519, or five-sixths of them all, were not over six months; and it now appears that on a scale twenty times larger than that given by Tardieu from his own experience, nearly two-thirds of the fœtal deaths induced by abortion were in from the third to the sixth month of pregnancy, the three periods included giving a much larger proportion than any others, and the last two of them being nearly identical. The extreme paucity shown by the above table in the first and ninth months, and the decrease in the seventh and eighth from those preceding, are worthy of remark. It is probable that the sudden increase may be attributable to mental reaction after the first shock occasioned by the absolute certainty of pregnancy, is past; and the subsequent decrease to the fact that in many attempted criminal abortions, during the latter months, children are born alive, the mother's courage then proving insufficient for infanticide, and its greater and more probable punishment."

My observation has shown me that criminal abortions are very frequent in the fourth week after conception. Many women, who are habitually regular to a day, are in the habit of using some drug or instrument if the menses delay a few days, and they have reason to suppose conception has taken place. They allege that they know by certain sensations that conception has occurred, as early as the third or fourth week. This habit is more frequent than has been supposed, as any physician will ascertain who seeks to investigate the matter.

In order to make complete the statistics of abortion, we should be able to present tables representing the average mortality of women who suffer that accident, both from criminal and other causes. "The results of abortion from natural causes," says Dr. Storer, "as obstetric disease, separate or in common, of mother, fœtus or membranes, or from a morbid habit consequent on its repetition, are much more than those following the average of labors at the full period. If the abortion be from accident, from external violence, mental shock, great constitutional disturbance from disease or poison, or even necessarily induced by the skillful physician in early pregnancy, the risks are worse. But if, taking into account the patient's constitution, her previous health, and the period of gestation, the abortion had been criminal, then the risks are infinitely increased." In 34 cases of criminal abortion reported by Tardieu, where the history was known, 22 were followed as a consequence by death, and only 12 were not. In 15 cases* necessarily induced by physicians, not one was fatal.

These meagre statistics are all I have been able to obtain relative to the consequences of abortion, to the mother. While I admit that the risks of a fatal result from criminal abortion brought about by other than skillful physicians, or even from diseased condition, are great, I cannot believe the result of abortion "necessarily induced by skillful physicians," is as fatal as Dr. Storer

---

* Annals de Hygiene, 1856, p. 147.

asserts. My observation and experience in this matter have been quite extensive, and I have been led to the conclusion that, if the operation is skillfully performed, the fatal results need not exceed one in a thousand.

I have not been able to find any statistics relative to the proportionate frequency of the various diseased conditions which result from abortion; but the information which I have received from my colleagues, and the various writers on diseases of women, as well as my own observations, lead me to place the consequences of abortion and their relative frequency in the following order:

1. Ulceration, erosion and congestion of the os uteri.
2. Premature and profuse menses.
3. Retroflexion and retroversion of the uterus.
4. Prolapsus uteri.
5. Ovarian disease.
6. Pelvic cellulitis.

The statistics relating to the *causes* of abortion, as well as those having reference to the various other matters connected with the subject, will be found under the appropriate chapters.

PART II.

CAUSES OF ABORTION.

CAUSES OF ABORTION.

The following Classification of the causes of Abortion is based upon the one used by Prof. Ludlam in his Lecture before the class of Hahnemann Medical College. The additions made by myself will be denoted by the initial H.

I. *Constitutional or Predisponent.*

1. Plethora.
2. Anæmia or Chlorosis.
3. The Scrofulous Diathesis.
4. Return of Menstrual Crisis.
5. *Zymotic Diseases.*
   - (a) Syphilis.
   - (b) Mercurialization.
   - (c) Variola.
   - (d) Scarlatina.
   - (e) Diphtheria.
   - (f) Cholera. (H.)

II. *Local or Organic.*

1. Malformation of ovum.
2. " of membrane.
3. *Placental Abnormalities.*
   - (a) Mal-location of (Placenta Previa.)
   - (b) Organic Disease of
   - (c) Detachment of
   - (d) Fatty Degeneration of
   - (e) Calcareous " of
   - (f) Hydatids. (H.)
   - (g) Moles. (H.)

III. *Reflex (Exciting.)*

1. *Centric.*
   - (a) Emotional—as Fright, Anger, Grief, etc.
   - (b) Direct blows upon the brain or spinal cord.
   - (c) Medicinal. (H.)
   - (d) Cerebro Spinal Meningitis.

2. *Concentric.*
- (a) Parotidean Irritation. (H.)
- (b) Thyroideal " (H.)
- (c) Thoracic " (H.)
- (d) Mammary "
- (e) Dental " (H.)
- (f) Gastric "
- (g) Rectal "
- (h) Vesical " also Renal. (H.)
- (i) Vaginal " (H.)
- (j) Hysterical "
- (k) Epilepsy.
- (l) Falls, jumping, blows, etc.
- (m) Functional and Organic Diseases of the Uterus. (H.)
- (n) Functional and Organic Diseases of the Ovaries. (H.)
- (o) Displacements of the Ovaries.
- (p) Death of Embryo.
- (q) Genital (Coitus.)
- (r) " (Instrumental.)

MEDICINAL.

Emmenagogues or Oxytoxics.

The following list of medicinal agents is necessarily imperfect, but I have placed therein the drugs which may, under certain circumstances, cause abortion. The list is not complete, as there are many medicines not named which have been supposed to cause that accident.

Apis mellifica.	Borax.	Podophyllum.
Actæa alba.	Bovista.	Quiniæ sulphas.
Aloes.	Cantharis.	Ruta graveolens
Asarum europeum.	Caulophyllum.	Sabina.
" canadense.	Cimicifuga.	Secale Cornutum.
Asclepias syriaca.	Decodon verticillatus.	Sanguinaria.
" incarnata.	Gossipium herbaceum.	Terebinth.
Aletris farinosa.	Ilex opaca.	Tanacetum vulgaris.
Baptisia tinctoria.	Mercurius.	Ustilago madis.

SECTION I.

PREDISPOSING CAUSES.

Plethora.

The first predisposing cause of Abortion mentioned in the foregoing table is *Plethora*; but it is oftener a cause of sterility, and will be again alluded to under that head. Obesity may cause abortion in several ways. The deposition of fat in the abdomen and pelvis will have the same effect as a tumor in those localities. In some cases of adiposis the action of the heart is manifestly impeded, and sometimes entirely arrested, by the deposition of fat in and around it. The same result may obtain in regard to the uterus, and prevent its *expansion beyond a certain point*, at which abortion would inevitably occur. Dr. Gardner also suggests that the deposit may "press down the uterus, pressing it into the pelvic strait, sometimes producing eversion, or so that the os uteri presses upon the sacrum, where by mere friction its surface becomes abraded, and profuse leucorrhœa ensues." Not only this, but prolapsus and retroversion may be thus caused.

Anæmia or Chlorosis.

Frequent abortions may cause anæmia; and this condition may in turn predispose to repeated abortions, the uterus not having sufficient "tone," or vitality, to retain the product of conception. Chlorotic women are as liable to abortion as those who are anæmic, and from the same cause, namely—a lack of vitality in the uterus, and a blood so impoverished that it is incapable of nourishing the fœtus.

Scrofula.

The *scrofulous diathesis*, by its influence upon the vital powers of the system, causing laxity of tissue, deficient nervous force, and depriving the blood of its normal constituents, is a powerful predisponent cause of loss of the ovum at any and all periods of gestation.

Return of the Menstrual Crisis.

This, as a cause of abortion, especially in the first months, has not been sufficiently appreciated. I have stated in the introduction to this work, that I consider it proper to define "Abortion," as the premature death and expulsion of the ovum at *any time after conception*, or before the end of the ninth month. I have also mentioned, in another place, the frequency of abortions in the *first* month.

The older accouchers paid much attention to the loss of the ovum shortly after impregnation. Married women who passed over a monthly period by a few days and then menstruated profusely were believed to have lost the ovum. This was called an effluxion, if it occurred before the tenth day, "because," as Smollie observes, "the embryo and secundines are not yet formed, and nothing but the liquid conception, or genitura, is dislodged. Tyler Smith thinks such cases are not uncommon, and the ovum is unobserved, not from its liquid condition, but because it is so little above the size of the unimpregnated ovum, as not to be visible in the discharges. An ovum of fourteen days has been described by Velpeau, and its size did not exceed three-eighths of an inch in diameter. In the expulsion of an ovum of an earlier date than this, the symptoms hardly differ from those of profuse menstruation. In cases of dysmenorrhœa—especially the pseudo-membranous variety, and in cases of profuse menstruation (habitual)—this loss of the ovum may frequently occur at a very early period. It has often been a matter of wonder to me that the uterus should so frequently resist the influence of a menstrual *nisus*, when abnormal in its character, or that a diseased uterus could so often bear the recurrence of the crisis when *normal*, without being irritated to such an extent as to throw off the impregnated ovum at such periods. (See *Ovarian Irritation*.)

ZYMOTIC DISEASES.

Syphilis.

Independent of the power which this blood-poison possesses of causing diseases of the uterus—as ulceration, etc., which will be mentioned in another place—syphilis seems to exercise a blighting power over the product of conception, in such a manner that it is liable to die in utero, of the same poison, and be expelled as a foreign body, at any period of pregnancy, or the membranes may become diseased from the pernicious influence of the malady: in both cases the disease may be transmitted by either or both parents: by the mother through the circulation, and by the father through the spermatozoa.

Writing of this disease, as a cause of sterility, Dr.

Gardner says: "In my opinion it is more often a cause of early abortion from an imperfect development of the ovum, than a cause of sterility, as it is difficult to decide whether the woman was ever impregnated or not. The menses are retained a few days over the usual period; there is finally a somewhat profuse discharge, accompanied by more pain than usual, and the passage of what are considered clots, but in them lies concealed the semi-developed ovum."

Mercurialization.

Mercury, in its various forms, may not only be an immediate cause of abortion, as in the cases of large doses of calomel, elsewhere alluded to, but the system may be so saturated with the poison that the blood and tissues are deprived, by its baleful influence, of the normal vitality necessary to carry on the process of gestation. The effects of this drug upon the organism are not unlike those of syphilis. It has long been remarked that women whose systems have been saturated with mercurial preparations were very liable to abortions, and this without any organic disease being discovered in the organs of generation.

Variola, etc.

Abortion may occur from the intense febrile orgasm, or from the congestive complications which take place during these disorders. All the *exanthematous* fevers predispose the system to take on such irritative conditions as may bring about miscarriage during or after their accession. The fœtus in utero has been known to be attacked with the variola, and be expelled either before or after its death.

According to Cazeaux, "confluent small-pox nearly always occasions abortion, and this is almost uniformly followed by the death of the mother."

Asiatic Cholera and Yellow Fever

are powerful zymotic poisons, and, like other diseases of this character, effect the fœtus through the medium of the circulation.

Dr. Pulte* states that during the cholera epidemic of 1848 abortions were very frequent, and were apparently caused by the imponderable poison pervading the atmosphere at that time. The same phenomena has been noticed during the prevalence of yellow fever in the large cities of the South.

Dr. Bouchut, in a quite recent work, mentions his observation of fifty-two cases of cholera in pregnant women, twenty-five of whom aborted in consequence of the disease.

SECTION II.

LOCAL CAUSES.

Abnormal conditions of the Ovum and its Appendages.

Dr. Whitehead says these are "so constantly associated with disease of the maternal organs as to lead to the suspicion that the mischief, in a great majority if not in all instances, originates in the latter." In my own practice, such complication has been almost invariably found in those cases wherein I have had an opportunity of making the necessary examination.

* Homœopathic Domestic Physician, 477.

The *fœtus* is liable to many diseases which may tend to its death, such as inflammation and dropsy of its venous cavities, dropsy of amnion, disease of the liver or kidneys, tubercular diseases, syphilis, and even small pox; diseases of the umbilical cord, knots upon the cord; strangulation, by the twisting of the cord about the neck of the child. I have seen cases in which the cord was twisted three and four times around the neck. Probably the diseased ovum excites the uterus to contraction before the actual death of the ovum has occurred.

In many cases of criminal abortion the injury done to the child by instruments, etc., is the real cause of the abortion. The membranes may not be ruptured or separated.

The impregnated ova may degenerate into *moles, hydatids,* or "*blighted ova.*"

Genuine moles are to be distinguished from certain fibrinous masses which are sometimes expelled from the uterus. These are called spurious moles, and a close examination will show their real character. Those moles which are the result of impregnation are of various kinds, consisting of different forms of degeneration of the membranes of the ovum. We can readily distinguish the varieties of moles depending on the carneous or fleshy, the hydatigenous, and the fatty, and other degenerations of the membranes. None of these cases can occur without conception. Many authors believed that fleshy moles might occur in nuns, and others presumed to be virgins, without the occurrence of intercourse. Percy believed that hydatids were independent animals, and that their production was compatible with the purest chastity. Dennison thought they sometimes originated in the uterus as independent formations, and Sir Charles Clarke was of opinion that uterine hydatids might exist apart from pregnancy. Madam Boivin and several other writers are in favor of the belief that this form of degenerated ovum may be retained for many months, or even years, after the ordinary date of labor. Tyler Smith is not aware that any case of this kind has been observed. I have known two cases in my own practice where a hydatid mass was not expelled until the *twelfth* month after the last appearance of the menses. Those who wish to investigate the nature of moles and hydatids can consult Tyler Smith.*

In the *carneous* moles there is an arrest of the usual symptoms of pregnancy, and the patient remains out of health. The ovum from the time of its death becomes to a great extent a foreign body, and is a source of irritation to the system generally. No increase of size takes place, so that at the fourth or fifth month the uterus may not be larger than it should be in the fifth or sixth week of normal pregnancy. The complexion is muddy and the breath fetid, with loss of appetite and digestion. Hemorrhage frequently occurs, but not very profuse.

The *hydatid* mole causes symptoms more strongly marked. The increase in size is often enormously rapid, so that at the fifth or sixth month the abdomen is as it should be at the end of pregnancy. The shape of the uterus is often quite different from that existing during natural pregnancy. There is absence of all fœtal movements and the sounds of the fœtal heart. After three or four months' suspension of the catamenia, there occurs a copious discharge of *water and blood*, resembling

---

* Lectures on Obstetrics.

red currant juice. This occurs irregularly and in variable quantities. The watery discharge is accompanied by pain, and appears to be caused by the breaking down of numbers of the larger **hydatids**. In a suspected case the discharges should be carefully examined, and of course the detection of a single hydatid renders the diagnosis certain. Excessive *flooding* often occurs, at frequent intervals, accompanied with discharges of masses of hydatids. The general health suffers profoundly, resulting in anæmia, dropsy, and even paralysis.

Fatty Deterioration of the Chorion and Placenta.

We are indebted to Dr. Robert Barnes for our knowledge of this frequent cause of abortion. This form of degeneration may affect the secundines at any time between the early weeks of pregnancy and the termination of gestation. Fatty degeneration may exist in the placenta as a post mortem change; that is, it may occur in utero after the death of the fœtus. It may happen also as the result of the transformation of effused fibrin in inflammatory disease of the placenta, or of a clot of blood in hemorrhagic effusion. Lastly, it may consist of the metamorphosis of portions of the maternal and fœtal structures of the placenta during the life of the fœtus. The latter pathological phenomenon is that which is of the chiefest importance in relation to abortion.

In a placenta affected with fatty degeneration, the lobes of the placenta are altered in appearance, some of them being of yellow fatty color, brittle and exsanguine, the rest presenting their ordinary characters. Examined more minutely, the tufts are found to be glistening, hard and tallowy, and not expanding when placed under water, as is the case with villi of healthy placentæ. The microscopical investigations of Dr. Hassall, show that the villi are studded with spherules and droplets of fatty matter and oil. The fatty material is found principally in the cells of the villi, and in the coats of its blood vessels, which vessels do not carry red globules when the degeneration exists to any extent. Dr. Barnes considers constitutional syphilis a frequent cause of this disease. Fatty degeneration causes abortion by destroying the vitality of the ovum; or owing to the friable condition of the placenta, partial separation may occur; or the partially degenerated blood vessels may burst and lead to placental apoplexy. For a full description of this disease of the placenta and chorion, see Tyler Smith's Lectures on Obstetrics, page 185.

Congestion of the Placenta

Leads to what is termed apoplectic effusion. Blood may be poured out either on the fœtal or external surface of the placenta. It may produce abortion in several ways. The loss of blood may deprive the fœtus of life; or the effusion may excite the separation of the ovum, and cause uterine contractions. In some cases the effused blood coagulates, its fluid portions are removed, and a fibrous mass remains without doing any great injury.

Inflammation of the Placenta (*Placentitis*).

In this disease effusion of lymph may occur, or the disease may pass on to hepatization, suppuration, or gangrene. Sometimes, when the inflammation affects the internal surface of the placenta, adhesions form between the placenta and the external surface of the ovum. In this way the placenta has been found adherent to the forehead or body of the fœtus. According to Professor Simpson, who wrote an elaborate memoir on this subject, the *symptoms* are obscure, consisting of pain in the uterus, near the site of the placenta, pains in the back and thighs, and general fever. Tyler Smith did not find, on stethoscopic examination, any modification of the uterine sounds in cases of suspected placentitis. The *causes* of placentitis are not very obvious, beyond mechanical injuries, and the great afflux of blood to the organ which occurs during pregnancy. Congestion and inflammation of the placenta are probably both common causes of abortion.

The *placenta* is liable to other diseases, as calcareous degeneration, tubercular deposits, and atrophy or hypertrophy. Sometimes, after the death of the fœtus, the placenta is still nourished, imperfectly, but still sufficiently to insure its retention, together with the dead ovum, for a considerable time.

Placenta Previa

May be the cause of abortion at any period of pregnancy. Of 378 cases mentioned by Whitehead, eight were from this cause. When the placenta is implanted on the *os uteri*, abortion is inevitable, and this almost invariably takes place before the fifth month. When only a small portion of it extends over the orifice, gestation may proceed to the seventh or eighth month, or even to the full period, without producing any great amount of danger to the process; but always, under such circumstances, separation takes place to some extent as the cervix expands, although premature expulsion is not an inevitable consequence. Whitehead mentions a case of implantation of placenta on the os uteri, and its partial separation, attended with hemorrhage before quickening, which he treated with favorable results.

SECTION III.

REFLEX CAUSES.

The *centric* causes of abortion are those which originate in the nervous centres, the brain or spinal cord, and act upon the uterus in a direct manner. They may be divided into Emotional, Physical, and Medicinal.

Emotional.—Fright, anger, joy, grief, and other mental influences, have been known to cause such disturbance in the organism, as to be a direct cause of the expulsion of the fœtus. Many cases are on record where the death of the fœtus has resulted from the perturbing effects of emotional shocks. Women have aborted immediately after hearing of the death of a beloved husband; or the gnawing canker of grief, shame, and remorse, has led to loss of the fœtus; and martyred women have aborted at the stake.

Physical.—Under this head we may enumerate direct blows upon the brain, or spinal cord, or intense congestion or paralysis of those organs. Some interesting experiments have been made, which have a bearing upon the centric causes of abortion. M. Serres divided the spinal cord in animals, after the commencement of parturition, and the *process was arrested*. In other experiments he excited abortion in animals by irritating the spinal marrow in the lumbar region. M. Brachet divided the cord in guinea pigs, between the twelfth and thirteenth dorsal vertebræ, after the commencement of labor, and everything but feeble contractions of the uterus was arrested, the animals dying in a few days undelivered. M. Segales made a section of the cord high up, without influencing the uterus, but the organ was paralyzed when the division was practiced low down. Cases are detailed by the above authors as occurring in the human subject, in which, in paralysis depending upon disease high up in the spinal marrow, uterine action was not interfered with; but was diminished or suspended altogether in cases of paraplegia—the result of injury or disease, low down in the cord. Dr. Simpson has lately made some experiments which go to negative the above. In his experiments, parturition is said to have occurred, notwithstanding the destruction of the lower portion of the spinal marrow. If such are Dr. Simpson's results, they will not prove the independence of the uterus of reflex action, since, from the connections of the greater and lesser splanchic nerves and the thoracic, abdominal and pelvic plexuses and ganglia, it is quite possible that the uterus may receive spinal fibres from the upper portion of the spinal marrow. The latter conditions may be caused by disease (idiopathic) or influences which we may term

Medicinal.—Certain medicines* undoubtedly cause abortion by their direct effect upon the brain or spinal cord. Those which produce congestive conditions are Quinine, Strychnine, and Ergot.

Quinine, according to Dr. Gardner, and several other observers, has been known to cause abortion, and they caution the practitioner against its use during pregnancy. The specific action of this drug upon the nervous centres is admitted by nearly all toxicologists. Dr. Brown-Sequard says it causes engorgement of the vessels of the brain and spinal cord. While engaged in an extensive practice, in a locality noted for malarious diseases, I found that abortions were frequent among the patients of my allopathic colleagues, who gave *quinine* in massive doses (5, 10, or 15 grains). The accident was supposed to be induced by the diseases for which it was administered. No such results occurred in my practice, although the cases under my care were as severe as any under allopathic treatment. Only a small portion of my patients were treated with *quinine*, and those to whom I was obliged to prescribe it took it in small doses, not exceeding one grain every two or three hours. I am fully satisfied that the miscarriages alluded to are caused by the *quinine*, and *not* by the chill or febrile paroxysm.

Strychnine has caused abortion. The tetanic spasms which occur in cases of poisoning by this drug seem detrimental to the life of the fœtus or the integrity of the uterine tissues. It causes congestion of the spinal cord and its membranes. The fœtus may not be expelled while the woman is under the influence of the poison, for in a frog, rendered tetanic by *strychnia*, the ova was not expelled during the tetanoid symptoms, but some days afterward, when the spasms had nearly disappeared.

Ergot.—This agent has been supposed by some to act upon the uterus *through* the cord. Tyler Smith says: "The ergot of rye passes into the blood, and affects the spinal centre, being specially directed to the lower part of the spinal marrow, and to that part of it in relation to the uterus." Others contend that it acts through the circulation directly upon the uterine tissues. Brown-Sequard, however, classes it among the medicines which cause congestion of the **vessels** of the spinal cord.

Carbonic acid, savin, aloes, alcohol, biborate of soda, and *ipecacuanha,* are supposed by Tyler Smith* and some others to act in a similar manner. This can, however, hardly be said of all the above agents. *Ipecac* acts upon the gastric nerves; *carbonic acid, alcohol,* and perhaps *borax,* may act through the cord, but *aloes* and **savin** appear to me to act in quite a different manner. The former seem to act by irritating the rectum, the latter as an irritant to the uterus through the medium of the blood, and rank with *tanacetum, turpentine, uva ursi, cantharis,* etc. If these medicines, however, are given in **massive doses, they may act** through both media.

There are other medicinal agents which may **act as** centric causes of abortion, namely, *gelseminum, caulophyllum, cimicifuga, cannabis indica,* and, perhaps, *gossipium. Gelseminum* probably causes **abortion by** paralyzing the lower portion of the spinal **cord. When** given in large quantities during **parturition, it arrests** that process. In small doses **it facilitates** labor, and increases the contractions of **the uterus.** The other remedies mentioned probably cause miscarriage by irritating the whole or a portion of the cord.

GALVANISM may be **applied** so as to act as a centric cause of abortion.

The state of the circulation affects the spinal centre in a very distinct manner. *Want* or *excess* of blood, or **materies** morbi in the circulation, act as a direct

---

* Lectures on Obstetrics, p. 264.

stimuli to the spinal centre, and in this way may induce abortion or premature labor. Certain diseases may act as predisponents, or exciting causes of abortion, by the influence which they exert *directly* upon the spinal centre. Among these diseases may be mentioned, cerebro-spinal-meningitis, myelitis, diphtheria, scarlatina, etc., examples of which will be mentioned in their appropriate places.

CEREBRO-SPINAL-MENINGITIS, OR SPOTTED FEVER. I find, on examining the reported cases of this disease, that it almost invariably causes abortion in those pregnant women who are its victims. This is what we might expect from a malady which strikes with fearful force the great nerve centres.

Dr. Black[*] reports a case of a woman aged thirty-one, mother of five children, and in the fourth month of her sixth pregnancy, of good constitution, who was taken with a prolonged chill, severe aching over the whole body, *and parturient pains*. On the afternoon of the same day had second chill, not so severe as first, succeeded by very high fever, and an *increase of pains. During the night she aborted. The fœtus appeared natural, the secundines passing without difficulty or hæmorrhage.*" This woman succumbed to the allopathic treatment on the tenth day.

Several cases have come under my observation, and to my knowledge, where this disease caused abortion. One peculiarity marks all the cases, namely, the abortion occurs shortly after the onset of the attack. I have been struck with the close similarity between the chill (rigor) and general symptoms of pain, prostration,

---

[*] American Journal of Medical Science, No. 98, p. 345.

etc., which attend the onset of both spotted fever and most cases of abortion.

SECTION IV.

CONCENTRIC OR REFLEX-SPINAL CAUSES OF ABORTION.

It will be well, before we enter into the consideration of the special concentric causes of abortion, if the physician or student refreshes his memory concerning the sympathetic relations of the uterus with other portions of the body, and the manner in which such relations are kept up. The uterus is in relation with the cerebral, spinal and ganglionic divisions of the nervous system, and possesses properties derived from each of these sources of motor power. The uterus is withdrawn from the direct influence of volition. The will has no direct power either to contract or dilate this organ. Labor may take place when cerebral paralysis exists—the will being entirely in abeyance, or abolished—as when under the influence of *chloroform, ether,* or *gelseminum.* But though not exerting any direct influence, volition may affect the uterus indirectly. (The direct influences—as emotions of various kinds—have been considered.) The efforts at "bearing down" during labor serve as an illustration of indirect cerebral influence; the abdominal muscles in this case stimulate the uterine by pressure. Since the brilliant discovery of the spinal system by Dr. Marshal Hall, that form of uterine action depending upon the spinal marrow now admits of clear comprehension, and is understood by all well-read physicians.

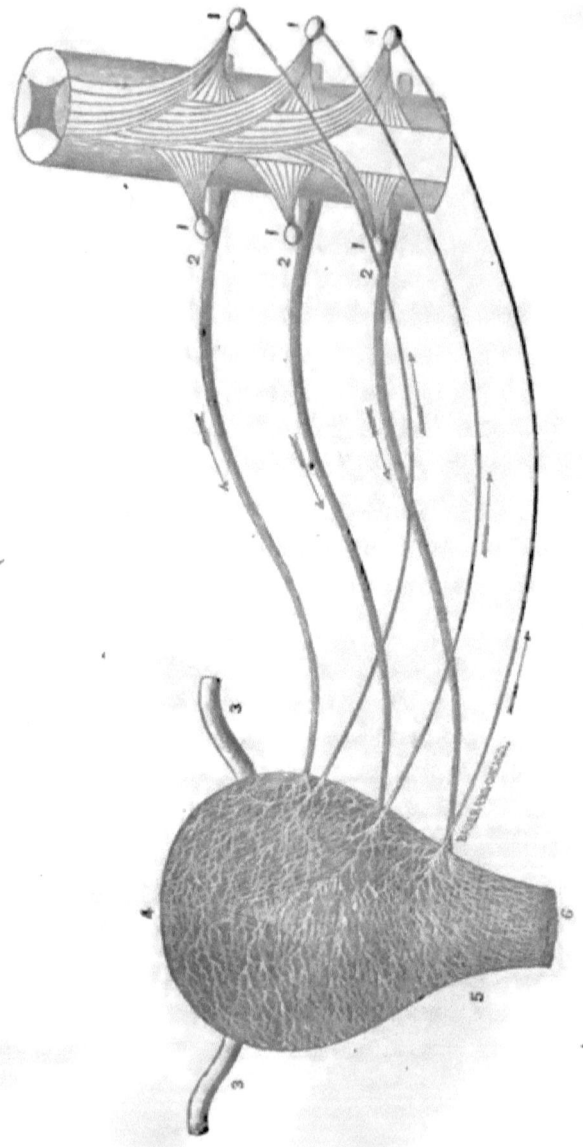

EXPLANATION OF PLATE.—1. Posterior or sensory **roots of** spinal nerves. 2. Anterior or motor ditto. 3. Fallopian tubes. 4. **Fundus** of the uterus. 5. **Cervix** uteri. 6. External **os** uteri. The *arrows* indicate the course of the **afferent and** efferent **currents.**

The best *resumé* of the *modus operandi* of this action, next to that of Dr. Tyler Smith,[*] is the clear elucidation by Dr. R. Ludlam,[†] in a valuable paper on the "Reflex Sympathies of the Uterus." From that paper I quote the following paragraphs, illustrated with the diagram kindly supplied to me by the author:

"The spinal nervous filaments supplied to the uterus are of two kinds—the motor and the sensory, or those which arise from the anterior and posterior columns of the medulla spinalis. The chief physiological peculiarity of these filaments is, that in case of the arteries and veins, their currents set in different directions—one toward, and the other out from, the central organ of the system. The sensory impression is that of general sensibility, and is afferent in its course—from the surface of the organ to which its filaments are distributed, and no matter how remotely it may be situated, to the spinal or cerebral centre. The motor impulse, or that which supplies the force that causes muscular contraction, is efferent in its course—from the cord or brain, or both, to the muscular tissue, upon which the motor nervous filaments are distributed.

"When you are told that every organ, and indeed every one of the bodily tissues, is supplied with, and is under the dominion of, the nerves, you will at once infer that the sensory and motor filaments must necessarily communicate with each other. This occurs either in the gray matter of the spinal cord, which is called its ganglion, or in that of the brain, where it forms the central ganglia. It is only necessary that the force or impression propagated to the sensitive filaments of the afferent nerves shall be conveyed to the gray or vesicular matter of the brain or the cord, when it is acted upon by some of their ganglia, modified and returned to the organ through the out-going conductor, the efferent or motor nerve.

[*] Lectures on Obstetrics, Page 263.
[†] North American Journal of Homœopathy, Vol. 13, Page 2.

"*Here is the whole philosophy of reflex action.* Every organ is connected with and under the control of a mass of gray neurine, which anatomists call a ganglion, no matter whether that collection of nerve vesicles be found in the brain, the spinal cord, or in the ganglia of the great sympathetic. Sensitive impressions telegraphed to, modified by, and returned from these various centres to the *same* or any *other* organ or organs, furnish all the detail of reflex action."

Our space will not permit an explanation of the physiological or healthy relations of the uterus to the cerebro-spinal system. Its pathological relations, however, come within the limits of our special subject. In order to be systematic we should classify these morbid relations under three heads, namely:

1. When the *sensory* current is reflected upon the womb, and causes various modifications of its healthy *sensibility*.

2. When the *motor* current is reflected upon the womb, causing abnormal or normal *mobility*.

3. When both currents are reflected in a way to modify the healthy *vascularity* of that organ.

From the *first* we may have hysteralgia, dysmenorrhœa, "uterine colics," super-sensitiveness of the uterus, etc.

From the *second*, uterine cramps, spasms, contractions, dilation and rigidity of the cervix, etc.

From the *third*, congestion, inflammation, and their sequelæ of leucorrhœa, ulceration, etc.

This treatise is, however, not intended to be sufficiently extensive for any attempt at such a complete classification of the causes of abortion. I shall, therefore, classify them under two heads.

1. Those irritations which, originating in the tissues, or within the cavity of the uterus, cause abnormal

influences to be reflected back upon itself, with sufficient force to cause abortion.

2. Those irritations which originate in *other* organs and are reflected upon the uterus, so as to cause the expulsion and death of the fœtus.

FIRST CLASS.

Parotidean.—In *Braithwaite's Retrospect*, Part xx. page 201, a case of abortion is mentioned, which seemed to have been caused by a metastasis of mumps. The patient was a lady aged twenty-five, who was attacked with cynanche-parotidea when advanced just beyond the third month of her third pregnancy. She had gone her full time in her previous pregnancies. After a day or two of vaginal discharge, uterine pains and hæmorrhage came on suddenly in the night and a fœtus was discharged. The hæmorrhage continuing, *ergot* was given, and fifteen hours after the birth of the fœtus the placenta was removed.

Dr. Salter, who reports the case, remarks, that we have abundant experience to show that the parotid glands, when diseased, have a relation to the testicles in the male and mammæ in the female. The mammary glands have a well-known sympathetic relation to the ovaries: and thus it may readily be supposed that in mumps an irritation may be communicated to the uterine system. On the other hand, the mammary glands may be affected, in metastasis of mumps, *through* the ovaries. It is difficult to decide which organ is primarily affected by the reflex influence originating in the parotids; but the analogy between the testes and the ovaries, and the sympathy of the breasts with the ovaries, go rather to support the hypothesis that the latter organs first receive the metastatic irritation.

Dr. Bedford and Scanzoni both make use of this reflex influence: the latter has founded upon it a method of inducing premature labor by irritation of the mammæ.

Thyroideal.—Cases are on record, and **some have come under my own observation, where abortion was** apparently **caused** by the application of *iodine* to an enlarged thyroid gland. All irritations of this gland should be avoided during pregnancy. I believe I was the first to call attention to the sympathetic relation of this gland to the reproductive organs, in a paper on that subject written several years since.* Subsequent investigations appear to substantiate my views on that subject.

Mammary.—Tyler Smith states that he has seen abortion caused by irritation of the mammary nerves. Instances of this are when abortions occur from prolonged lactation during pregnancy. That it is not mere weakness or exhaustion in **some of these cases is** proved by the fact that the mammary secretion may cease upon the occurrence of impregnation, but that a plentiful supply of milk returns after the recurrence of abortion. **Blisters** or sinapisms, or even hot **fomentations to the** breasts, may irritate the **pregnant** uterus. It is well known that contractions of the womb are excited, after labor, from irritation of the mammæ.

Gastric.—Although it is surprising what an amount of nausea and vomiting the uterus will bear without being excited to expel its contents, yet there are many cases recorded in which abortion has been apparently brought on by excessive vomiting during pregnancy. I have known abortion to be caused by the excessive

---

* North American Journal of Homœopathy, **Vol. 12, page 375.**

vomiting induced by lobelia. It is true that in some cases the mechanical irritation may be the main cause, but there are many in which it undoubtedly occurs as a reflex pathological phenomenon.

Dental.—Irritation of the trifacial nerves may produce abortion. This happens sometimes from the irritation of cutting the wisdom teeth, the extraction of a decayed tooth, or the irritation of a constant odontalgia. It is advised by many of the older obstetric writers to avoid the extraction of teeth in a pregnant woman.

Renal.—Acute nephritis, the passage of calculi, or irritation of the kidneys in albuminuria, are said to be sometime causes of abortion.

Vesical.—Cystitis—idiopathic, or from the poisonous action of drugs—may be a cause of abortion. *Turpentine, cantharis,* and other medicines, may induce such irritation of the bladder as to bring on miscarriage; so, also, the presence of stone in the bladder.

Rectal.—The production of abortion by the irritation of the rectal nerves is a well-recognized occurrence. It may happen from hœmorrhoidal inflammation, the irritation of ascarides, the action of violent purgatives, diarrhœa, dysentery, or the opposite condition of excessive constipation, stimulating enemata, etc. Whitehead, under the head of "Functional Impediments of the Bowels," mentions many instances of abortion from the irritation of retained fœcal matter. He says: "The symptoms are distension and tenderness of the abdomen, commonly attributed by the patients to flatulence, of which they are constantly endeavoring to relieve themselves by eructation. The abdomen is sometimes as large, under these circumstances, in the middle, as it should be at the end of pregnancy. There is a constant inclination to relieve the bowels, the evacuations, which are thin and scanty, being accompanied by violent *straining* efforts. This action is not long in being extended to the uterus, which becomes affected with pains of an intermittent and expulsive character, creating the belief that abortion is about **to happen.** The **real** cause of these disturbances is accumulation of fœces in the third turn of the colon, accompanied with flatulent distension of that portion of the bowel immediately above **the seat of obstruction.** This cause of abortion does **not generally occur before the fourth** month, **for not until** that time does the womb attain sufficient size to act as an impediment to the passage of fœcal matter through the rectum. The hæmorrhage in this description of cases is apt to be profuse, and restrained with difficulty." Fœcal accumulations may also **cause a**bortion, by producing that displacement of the uterus known as retroversion, a condition treated of in another place. *Aloes*, *podophyllum*, *mercury*, and their analogues, may cause abortion by **the rectal irri**tation **they** produce.

Vaginal.—Acute vaginitis, gonorrhœal or idiopathic, will sometimes **cause** abortion. Mechanical irritation of **the vagina,** by plugging, coition, **or ill-**fitting pessaries, **or irritating** injections, may have the same effect. Vaginismus, a painful spasmodic affection, which has attracted much attention of **late, is** not only a frequent cause **of** sterility, but of abortion. Dr. Helmuth* **has** given **us** a vivid example **of** this condition, in **a** recent article. The induction of abortion by means of the *colpeurynter* will be considered in another place.

Ovarian.—Any irritation or excitement **of** the ovaries is reflected upon the uterus, mammæ, **or** thyroid gland.

***See** Western Homœopathic Observer, Vol. 1.

In the unimpregnated state, the uterus generally receives the reflex ovarian influence; when it occurs during pregnancy it is an abnormal phenomenon, and is liable to cause miscarriage. If this influence is received by the mammæ or thyroid, the uterus is left to go on with its normal development. In many instances the simple occurrence of the menstrual nidus has been sufficient to cause abortion. This is one form of that disease called "habitual abortion." The uterus must first be in a debilitated, irritable condition, in order to be seriously affected by this influence. The ovarian irritation may be perfectly normal all the time, or it may be abnormal, consequent on diseased conditions, or the toxical effects of drugs. Thus if *cantharides, camabis indica, apis mel*, or similarly acting drugs, be given in pathogenetic doses, they may set up an irritation in the ovaries which shall be reflected upon the uterus with such force as to cause it to take on diastaltic action. Certain diseases of the ovaries have the same effect, namely — inflammation, congestion, suppuration, etc. Those diseases which can act as direct or mechanical irritants to the uterus, are enlargements, from cystic and other growth, dropsy, etc.

SECOND CLASS.

Uterine.—This class includes the following, namely:
(*a*) Diseases of the uterus, functional and organic.
(*b*) Displacements of the uterus.
(*c*) Death of the fœtus.

The uterus is endowed with what is termed *peristaltic action*—a peculiar vermicular motion or contraction, which is called by Tyler Smith, "ganglionic motor action." "When any part of a muscular organ, supplied in whole or in part by the ganglionic system of nerves, is irritated, the contraction which ensues generally spreads in a vermicular manner to a distance from the point of irritation, and continues for some time after the exciting cause is removed. The uterus is eminently endowed with the peristaltic form of contraction. When one point of the uterus is stimulated, through the abdominal parietes, or by the introduction of the hand into the uterus, the contraction excited extends to the whole organ." Dr. Smith has seen this action occur after death, in animals, and he asserts that the uterus seems capable of expelling the foetus *by peristaltic action alone.* In cases of paraplegia from disease of the lower part of the spinal marrow, or in animals reduced to the same state by experiment, the peristaltic action is the chief power remaining in the uterus. In such cases delivery has been effected in an imperfect manner by the peristaltic action of the uterus or by the application of galvanism to the organ. It is not stated, however, how much of the spinal marrow must be destroyed before the reflex or disastaltic action of the uterus ceases. It is not probable that this peristaltic action can exist to any extent, unconnected with the reflex spinal influence. In cases of labor and abortion, the two forms of action exist in combination. The motor nerves of the uterus are in relation with the mammary, pubic, rectal, pneumogastric, ovarian, vaginal, and the nerves of the os and cervix uteri, as incident excitor nerves. There can be no doubt that in an organ thus subject to reflex action, its own nerves are excitors, and that in all contractions of the uterus excited by irritation of its external surface, or of the os and cervix, by disease or otherwise, the uterine actions are both reflex and peristaltic. It is indeed a question if any pure spinal fibres reach or proceed from the uterus unmixed with fibres from the ganglionic. Any pathological condition of the uterus, which is capable of exciting these reflex-spinal and ganglionic-motor actions, may be a cause of abortion at any period of pregnancy.

Functional diseases of the uterus, are those in which no change in structure has yet appeared, although the condition existing may lead to such lesion. The older writers laid much stress on the opposite states, which they termed *rigidity* and *laxity* of the muscular tissue of the womb. We cannot dispute the fact that such a condition may exist as idiopathic affections, *i. e.*, depending upon the same *general* state of muscular fibre. But it may obtain from purely local causes, namely: an exhausting discharge from the mucous tissues of the organ itself—as functional leucorrhœa. But as *leucorrhœa* may be functional, and also dependent on organic changes in the uterine tissues, we will proceed to consider that disease, as it is the connecting link between the two classes of uterine diseases.

SECTION V.

FUNCTIONAL DISEASES OF THE UTERUS.

Congestion of the Uterine Circulation.—According to Whitehead this appears to prevail as the immediate cause of abortion in one out of every twenty-five cases. He believes the average is even greater. Those in whom the venous capillary function is naturally below par, indicating predisposition to local congestion, are most frequently the subjects of it. The *symptoms* which usually manifest themselves after the period of quickening, from the end of the fourth to the eighth month of pregnancy, are—immoderate and painful distension of the abdomen, generally attributed by the patient to the accumulation of wind in the bowels; a pulsative movement extending over the whole cavity; **sense of weight and bearing down**; intermittent pains in the loins, like those of labor, and occasionally escape of blood from the vagina. There is also distension of the pubic, spermatic, hæmorrhoidal, and all the pelvic veins, and sometimes those of the lower extremities. *On examination*, the vagina is found hot and turgid, and the cervix uteri tumid and varicose. When this state is allowed to exist for a length of time, local phlebitis may take place, resulting in varicose ulceration of the cervix; or the inflammation may extend through the entire organ, **and eventually to the uterine peritoneum, ending in effusion, etc.**

Leucorrhœa.

The frequency of leucorrhœa during pregnancy, and the many unpleasant symptoms to which it gives rise, should lead us to study more closely the connection between that disorder and abortion.

The valuable monograph of Dr. Tyler Smith on Leucorrhœa leaves but little to be desired relative to the true pathology of the abnormal discharges which go under that name. The limits of this work will not permit me to give more than a cursory glance at the divisions of leucorrhœa; our main purpose is to consider the influence it has on the causes of utero-gestation. But the *treatment* of this disease will in time be so intimately connected with a true understanding of the nature and locality of the discharge, that we cannot forbear giving some general idea of the classification of the varieties of this malady.

"All pathology has its basis in physiology. The demonstration of two very differently organized surfaces in the vagina, and in the canal of the cervix uteri, with the existence of two very distinct forms of secretion, naturally lead us to the consideration of two principal forms of leucorrhœa. But at this point it may be well to revert for a moment to the special difference which exists between the vagina and the cervical canal. The lining membrane of the vagina approaches in organization to the skin; it is covered by a thick layer of scaly epithelium; it contains in the greater part of its surface few, if any, mucous follicles or glands; its secretion is *acid*, consisting entirely of plasma and epithelium, and the chief object of the secretion is the lubrication of the surface upon which it is formed.

"On the other hand, the lining of the canal of the cervix, is a true mucous membrane: it is covered in great part by cylinder epithelium; it abounds with immense numbers of mucous follicles, having a special arrangement; it pours forth a true mucous secretion, *alkaline* in character, and consisting of mucous corpuscles and plasma, with little or no epithelium, and this secretion has special uses to perform in the unimpregnated state, and in pregnancy, and parturition."*

We here have presented the anatomico-physiological character of the vagina, and canal of the cervix uteri. So long as the secretions from these surfaces remain within physiological limits, no disease is present: but the moment these secretions become *abnormal in quantity or quality*, the result is *Leucorrhœa*.

Leucorrhœa admits of the same divisions as set forth in the above quotation. The *first* and most important is the *Mucous* variety, consisting chiefly of mucus-cor-

* Tyler Smith on Leucorrhœa.

puscles and plasma, and secreted chiefly by the follicular canal of the cervix. The *second* is the *Epithelial* variety, in which the discharge is vaginal, or is secreted by the vaginal portion of the os and cervix, and consists, for the most part, of scaly epithelium and its *debris*.

These two varieties may, of course, exist in various degrees of combination. Sometimes the one and sometimes the other preponderates, or is the original affection. The old division of *uterine* leucorrhœa, as arising from the cavity of the *fundus*, is now obsolete—such discharge rarely occurs. In certain cases of menorrhagia, the periodical sanguineous discharge is converted into a constant colored discharge, in which may appear some mucus, but hardly enough to constitute a leucorrhœa.

Cervical or Mucous Leucorrhœa.—This most common form of leucorrhœa is, when simple and uncomplicated, the result of a morbid activity of the glandular cervix. Instead of the discharge of the plug of mucus at the catamenial period, a constant discharge is set up. The glandular portion of the canal of the cervix is the chief source of the discharge; it is a special secretion, elaborated by those glands.

In recent cases of cervical leucorrhœa, when the disorder consists merely of a hyper-secretion of the mucous follicles, without any manifest lesion of structure, the cervical discharge is found (on examination with the speculum) hanging at the os uteri, or adhering to its vaginal portion, and is almost always viscid and transparent. It may be drawn out in long tenacious threads of the utmost clearness, unless in course of pregnancy, or abnormal state of the vagina, when it is rendered opaque by the *acid* vaginal mucus. This string of mucus sometimes extends the whole length of the vagina, and even extends from that passage. This secretion is always *alkaline*, in contradistinction to the vaginal secretions which are *acid*.

In severe or chronic cases of this form of leucorrhœa, the alkaline cervical mucus is mixed with pus and blood, owing to the irritable and more deeply diseased condition of the glands of the canal of the cervix. In some cases, the exudation of blood from the canal of the cervix is so constant that it is apt to be mistaken for menorrhagia. In other cases of cervical leucorrhœa the secretion is so profuse and watery that the traces of viscidity are nearly lost. Instead of the mucus and plasma, a watery serum is poured out in large quantities.

The quantity of mucus or serum lost in cases of cervical leucorrhœa is often so considerable as to prove a serious drain to the constitution, and set up functional or more serious disorders in different parts of the body. The serous secretion in particular is often a source of great debility. Patients suffering from either of the above forms may become hectic from purulent secretion and absorption, or rendered anæmic from the loss of blood. The *symptoms* arising from cervical leucorrhœa are numerous and changeable, dependent on the amount of local irritation, functional derangement of other organs, or the loss of tone in the muscular or nervous system.

Sequelæ of Cervical Leucorrhœa.—It is well known by the reading men of the medical profession, that writers on uterine pathology are divide into antagonistic schools, namely, (1) those who, under the leadership of Bennet, believe nearly all abnormal discharges from the uterus are the result of inflammation and ulceration, and (2) those who adopt the theory of Tyler

Smith, that the ulceration is the result of abnormal discharges from the cervical canal and its glandular apparatus.

When these conditions occur as actual sequelæ of cervical leucorrhœa, the following is their order of appearance. I quote from Dr. Tyler Smith.

"By observing cases of mucous or cervical leucorrhœa, under every variety of circumstances, we may obtain a tolerably correct knowledge of the different stages of the disease, and we may learn the order in which its sequelæ makes its appearance when it is allowed to run its course unchecked for a considerable time. In the first place there is simply an increase of the secretion of the cervical mucus. Instead of the formation of the plug after each monthly period, there is a constant escape of thick mucus from the os uteri. But in this phase of the disorder there is little constitutional or local disturbance. The size of the os and cervix is not increased, and the surface of the os remains quite natural, both as regards volume and color. After a time the os uteri gapes; there is relaxation of the cervix, the upper part of the vagina loses its tone, and some amount of prolapsus generally occurs. With this the ring of superficial redness slowly passes on to the destruction of epithelium; then the loss of the villi takes place, and the formation of the granular surface upon their base occurs. The whole of the os and cervix now becomes swollen and turgid, induration commences, and fibrinous deposit in the substance of the cervix frequently takes place. The sensibility of the different portions of the utero-vaginal canal varies greatly in different cases. In some the abraded or hypertrophied os uteri is exquisitely tender, while in others its sensibility is little, if at all, increased. In some cases of leucorrhœa, in which abrasion occurs, the whole of the os uteri and the cervix hangs into the vagina, completely denuded of its integumentary covering, but there is no great enlargement of the parts. In others there is considerable hypertrophy without any destruction of epithelium or loss of surface."

All the conditions above mentioned as being caused by *cervical* leucorrhœa may arise from other causes. Dr. Bennet insists that they arise from *inflammation* of the cervix.

It is of considerable importance to the practical physician whether these conditions are primary or secondary effects: the selection of the remedy depends considerably upon the theory adopted. The controversy is yet undecided, and probably will not be until personal asperity and prejudice is laid aside, and all seek conscientiously for truth. As in nearly all other controversies, neither party has all the truth on his side. I am inclined to the belief that either leucorrhœa or inflammation may be the primary affection. An inflammation of the cervix, or an ulcer thereon, may cause cervical leucorrhœa, and *vice versa*. The best we can do, in the present state of our knowledge, is to judge from the history of the case, and a careful examination with the speculum, which has the priority.

Vaginal or Epithelial Leucorrhœa.—This discharge has its origin in the muco-cutaneous lining of the vagina and the portion of this membrane reflected on the external surface of the cervix to the margin of the os uteri. In strictly vaginal leucorrhœa there may be no discharge whatever issuing from the canal of the cervix, and in some cases the secretion of the cervix seems almost suspended, the os uteri appearing drier than natural, and no mucus being visible between the labia uteri. In others the cervical glands are excited by the condition of the vagina, and secrete copiously a mixed epithelial and mucous leucorrhœa, from the union of the two kinds of discharge. The discharge in vaginal leucorrhœa may arise, chiefly, either from the lower part of the vaginal membrane or from that part which is reflected upon the cervix, but in severe cases the whole surface of the vagina is involved. In simple acute cases of vaginal leucorrhœa the discharge is epithelial, made up of imperfect and perfect scales. In severe and chronic cases, *pus* is mixed with the epithelial matter; for the villi become affected and the pus is formed upon the sub epithelial or villous surface. In some cases portions of the vaginal surface may be so abraded that blood globules escape and mix with the other constituents of the vaginal discharge. In one form of vaginal leucorrhœa but little fluid discharge appears, but the vaginal walls are coated over with a white membrane, which may be detached in large shreds or pieces, composed of epithelium in which the parchment-like arrangement of the scales is perfectly preserved. These laminæ frequently have upon them marks of the rugæ of the vagina, and their under surfaces are rough from the indentations of the vaginal papillæ. This may be termed a *membranous* form of leucorrhœa, and occurs oftenest in cases of pregnancy. The vagina may be attacked with the diphtheritic poison, and secrete a membrane having that character. In some cases of vaginal leucorrhœa, the irritation is intense and annoying, assuming the form of *pruritus vulvæ*. This latter symptom is, however, oftener caused by an aphthous-inflammation of the lining membrane of the vagina and vulva. I have known cases in which abortion appeared to be caused by this symptom alone, the severe reflex irritation being sufficient to set up expulsive action in the uterus.

The frequency of leucorrhœa during pregnancy has been alluded to. Whitehead found, in 2,000 cases 1,116 in which leucorrhœa was present. Out of the 2,000 cases, 747 had abortions, and of this latter number, only 172 cases of abortion could be assigned to *specified* causes—leaving 575 cases out of 1,116 having leucorrhœa. This is a large percentage—larger than we ever meet in general practice. In many of these cases, perhaps a majority, the discharge was probably due to ulceration.

Gonorrhœa.

This should not be overlooked when we are considering the causes of abortion. The uterus is more commonly affected by the gonorrhœal poison than has been supposed. Whitehead asserts that this disease more commonly affects the uterus than the vagina. This opinion is at variance with what has usually been taught. That writer contends that the gonorrhœal virus, from physiological causes, is liable to be carried immediately to the highest part of the canal, and forcibly projected upon the lowest extremity of the uterus, which organ also, at this juncture, is in a state eminently calculated speedily to absorb it; besides, the normal secretion of the vagina possesses properties which are capable, to a certain extent, of destroying or materially modifying the virulency of the poison, and of thus protecting the vaginal surface from its immediate influence. The urethral orifice, however, does not seem to be provided with this protection to any degree, and is therefore much more susceptible to the action of specific inoculation. In nine unimpregnated women affected with gonorrhœa, seven had inflammation, with abrasion of the os uteri, and in the remaining two, the upper vaginal surface and urethra were affected.

Gonorrhœa first affects the uterus by causing superficial inflammation of the lips of the os and the commencement of the internal cervix. The inflammation seems to affect principally the small mucous follicles with which the surface is closely studded. "A small red patch is first perceived; sometimes there are two or three isolated spots which extend and soon run together, forming one patch, of variable size in different cases and in different stages of the complaint, and generally of irregular shape. On removing the thick secretion with which this is covered, the surface appears to consist of minute granules, equally dispersed over every part of it; the abrasion is bounded by a margin not very distinctly defined, running imperceptibly into the erysipelatous redness which surrounds the sore; this extends to some distance upon the cervix, the whole of which is more or less *tumid*, but not painful to the touch." In plethoric women the symptoms are often very violent, the inflammation being severe, and the ulceration of an irritable character, throwing off large quantities of pus and often causing much fever.

If the infection be contracted during pregnancy, abortion is liable to take place during the acute stage of the complaint from the intense uterine irritation. If the disease existed previous to gestation, it may have caused chronic endo-uteritis, or an ulcerated condition of the cervix, both of which conditions are quite serious, and are likely to be cause of abortion at any period of pregnancy.

It is Whitehead's opinion that gonorrhœal affections in women are rarely cured; that they are frequently the cause of induration, fissured ulcer, and chronic inflammation of the deeper seated tissues.

SECTION VI.

ORGANIC DISEASES OF THE UTERUS AND CERVIX.

Ulceration of the Cervix.—Whitehead refers to a table containing the records of 400 cases of abortion and threatened abortion, " In all which, disease of the uterus was an accompanying condition, and for which no other cause could be assigned for the disturbance complained of, the average occurrence of the superficial granulating ulcer, or of diffuse inflammation of the cervix, amounted to 26 in every hundred. In the majority of these, the event happened between the middle of the sixth and the middle of the ninth month. In some, however, the symptoms commenced earlier."

This would seem to indicate that ulceration of the cervix was better tolerated in the earlier, than in the later months of pregnancy. In several cases, recorded by Bennet,* the abortions, or threatened abortions, occurred as often in the first, as in the last three months of gestation.

Ulceration of the cervix may be considered as one of the chief causes of abortion. Bennet thus sums up his experience in the matter:

"Abortion is often occasioned by inflammatory ulceration of the cervix, and likewise often occasions it. In the latter case, abortion occurs accidentally, under the influence of some of its generally recognized causes, and leaves behind a morbid state of the cervix and its cavity. Local disease of this nature may follow an

---

* Diseases of the Uterus.

abortion of the simplest kind, one from which the patient rallies in a few days, although it is more generally the result of those that are accompanied by inflammatory and hæmorrhagic symptoms. Ulcerative disease of the cervix, when once established, from whatever cause, is itself a frequent cause of abortion. When abortion is the result of the actual existence of the inflammatory disease, it may be produced in various ways. The vitality of the womb may be so modified in the earlier stage of pregnancy, by the existence of the disease, that the fœtal germ dies, in which case it is expelled along with the membranes, or it is partly or entirely absorbed, the membranes continuing to enlarge for some months, and being eventually expelled under the form of a mole, or false conception; or the pregnancy may advance to a farther period, until the third or fourth month, when the womb becoming too irritable, or being unable to develop itself, or the fœtus dying, the membranes separate, flooding ensues, and the contents of the uterus are expelled. At a later stage still, when the muscular structure of the womb is more fully developed, the presence of inflammation at its mouth may bring on strong reflex action, and occasion premature confinement, independently of any disease of the child or of its membrane.

"Abortions, no doubt, frequently occur under the influence of accidental causes alone, and of constitutional cachexia, such as scrofula and syphilis, without there being any local disease of the cervix. It may, however, be laid down as a rule, that a great majority of abortions which are preceded or followed by morbid symptoms, and of those which occur spontaneously, without any evident cause, and in the absence of uterine tumor or constitutional cachexiæ, are occasioned by inflammatory disease of the cervix. It may also be considered as all but certain, that inflammatory and ulcerative disease of the cervix exists when abortions quickly succeed one another, and when a female does not seem able to carry the product of impregnation to the full time."

According to the same author, "inflammatory ulceration, during the pregnant state, is by no means *necessarily* followed by abortion." Gestation may go on till full time, when no remedial treatment is resorted to; or the ulceration may be treated by local applications, and cured, without interfering with the course of pregnancy.

Bennet also says that "*instrumental* examination of females [women?] laboring under inflammatory ulceration of the cervix, during pregnancy, is unattended with any risk, either to the mother or to the fœtus, and it is absolutely necessary, in order not only to fully recognize the disease, but also to treat it."

This statement I have often verified in practice. The speculum should be generally resorted to in cases of threatened abortion.

"The *simple granulating ulcer*," says Whitehead, "may be confined to one labium only, the other being perfectly normal. More frequently, however, it implicates both at the same time, extending to some distance up the external cervix, and passing more or less within the orifice, which often appears to be the part most severely affected. The whole cervix is in a state of hypertrophy, and considerably softened, with the exception of the inflamed crust upon which the ulcer is situated.

"Upon *tactile* examination, the whole lower part of the uterus is found to be altered in form: the lips are elongated and flabby, and the orifice open. The ulcer presents a flattened, velvety surface, with a raised cord-like external boundary, which the practiced touch will be able to detect without difficulty. When viewed through the SPECULUM, the whole cervix—unless it be unusually large—will readily fall within the upper aperture of the instrument. The diseased surface, when both labia are implicated, appears irregularly circular, about the size of a shilling—larger or smaller—of a bright red color, and covered over with a coating of muco-pus: this being removed by a piece of lint, to which a portion of ropy mucus often adheres, derived from the central orifice, the granulations are brought palpably to view. * * * The granulating ulcer is most commonly observed in women of the sanguino-lymphatic temperament, lax fibre, and feeble circulating powers. When met with in primapara, the first indications of its existence are noticed before the period of quickening, often as early as the second or third month. It may exist for years, and during several pregnancies, without causing abortion." (See plate.)

West[*] describes ulceration of the cervix in a different manner:

"They are for the most part mere superficial abrasions of the epithelium investing the lips of the os uteri, whose abraded surface is of a vivid red color, and finely granular. In other cases, in which the absence of epithelium is less complete, the surface seems beset by a number of minute, superficial, aphthous ulcerations, between which the tissue appears healthy, or slightly redder than natural. The ulcerations of the os uteri seldom or never present an excavated appearance with raised edges, as ulcers of other parts often do, but either their surface is smooth, or it projects a little beyond the surface of the surrounding tissue. They are usually, but not constantly, of greater extent on the posterior, than on the anterior lip, are sometimes confined to the former, but very rarely indeed limited to the latter. They appear to commence at the inner margin of the os uteri, whence they extend outward, and sometimes, though by no means invariably, the short extent of the canal of the cervix uteri which can be brought into view by the speculum, appears denuded of its epithelium."

[*] Diseases of Women, page 97.

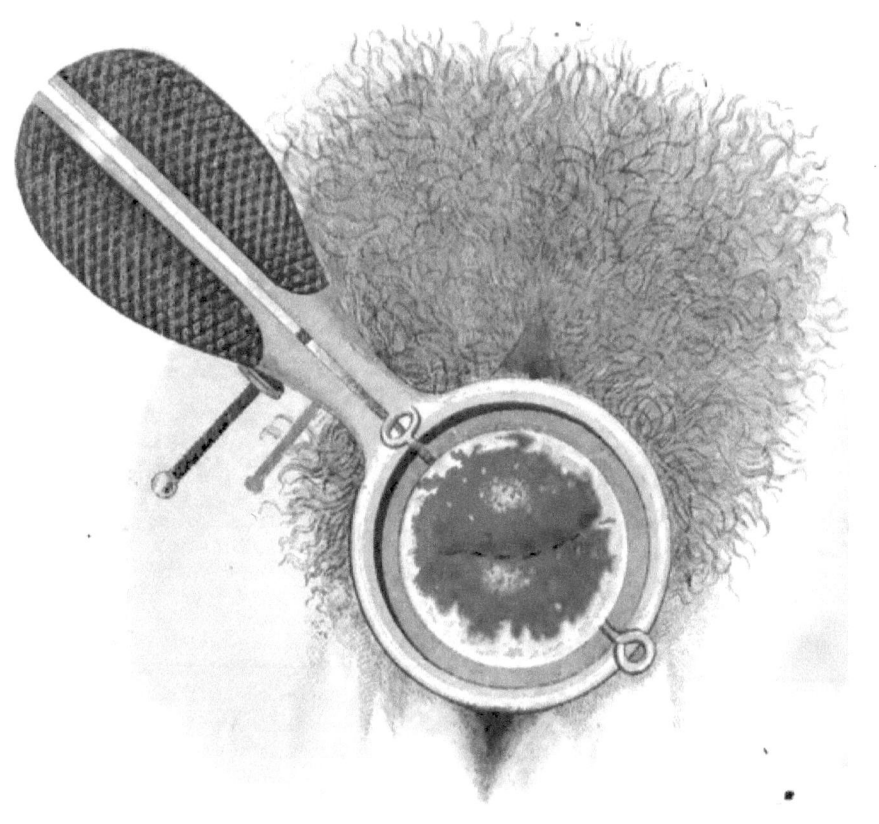

Other writers vary as much in their description of ulceration of the os and cervix, as Bennet says, "without necessity or advantage."

Writing of this lesion the same writer says:

"An ulceration occupying the cervix uteri may present the various modifications which suppurating surfaces offer in any other part of the body, from the minute granulations of a slight abrasion, to the livid vegetations of an unhealthy sore; but these modifications of the ulceration require, in reality, no division or classification."

The student who would acquaint himself with the various local and constitutional *symptoms* caused by these ulcerations, may refer to such works as West, Meigs, Scanzoni, Simpson, and particularly Bennet.

To the Allopathic school, who attack all ulcerations of the cervix uteri with caustics or constitutional remedies—tonics and alteratives—the division of ulceration into different forms and species, may be of no practical use or benefit; but to the homœopathician, these classifications, or minute descriptions, may be of great value in the selection of the specific remedy, especially when our Materia Medica shall be perfected, particularly the pathological symptoms thereof.

After consulting all the more recent writers who describe ulceration of the cervix uteri, and from the results of my own limited observation, I know no better manner in which to classify these lesions than the one adopted by Whitehead—namely:

Superficial erosion,   *Fissured ulceration,*
Varicose ulceration,   *Follicular ulceration.*

Bennet inadvertently sanctions this arrangement, for he mentions—"Abrasion, or Excoriation"—"Ulceration, with Fungous Granulations"—and "Varicose condition" of the parts—" Aphthæ, or Ulcerated Mucous Follicles"—and even a condition when "The lips of the os uteri are very much hypertrophied and indurated * * separated by a deep fissure, and the ulcerated surface, which is situated *deeply between them*, can only be discovered with the eye, on their being separated by a bivalve speculum."

We will now examine the peculiar characteristics of the different forms of ulceration: the simple *erosion*, or *granulating ulcer*, has already been described.

Varicose Ulcer.—This variety, according to Whitehead, is liable to cause abortion in the latter two or three months of pregnancy. It prevails in about six or eight cases out of every hundred, and is often difficult of cure before delivery. It is generally met with in women of the bilious temperament and hard fibre, who have been subject to piles and profuse menstrual discharges, and to derangement of the biliary organs. The premonitory condition of the parts consists in a hardened and hypertrophied state of the cervix, which is traversed in various directions by a number of tortuous dark colored trunks, about the thickness of a probe, or goose quill, raised upon the surrounding surface. Larger and more prominent points are here and there noticed, indicating the situation of inosculation of one branch with another; and generally, at one of these points, the ulcerative process is set up, which soon extends through the coats of the vessel, and hæmorrhage follows immediately. The ulcer, which is not long after in being developed, presents an uneven livid aspect, with irregular margins, near which a few tortuous vessels may be seen ramifying: it now secretes a quantity of pus, and often has small dark clots of blood or fibrin, the size of a pin's head, lying loose upon the surface.

It usually occupies but one labium, the anterior more frequently, but often the whole circumference of the cervix is implicated. Sometimes the inflammation extends to the body of the uterus, as an acute phlebitis, and thus acquires a serious importance. The *symptoms* indicating the existence of varicose ulceration during pregnancy, are bearing-down pains of an intermittent character, similar to those of the first stage of labor, aching of the loins and along the thighs, irritable bladder, with inability to retain the urine the usual length of time, disorded digestion, sickness and headache, languor, and a vaginal discharge of a white glairy mucus at first, which becomes brown, then bloody, and finally purulent. Bennet says this form of ulceration is generally attended with fungous granulations, large, livid, and bleeding at the slightest touch. Whitehead says it causes induration of the cervix, and often degenerates into the

Fissured Ulcer.—This form of uterine disease, according to Whitehead, is perhaps equally prevalent with the superficial variety, and is much more difficult of cure, on account of the extent to which the subjacent textures are implicated. It is found to exist in twenty to twenty-four out of every hundred cases of abortion, not resulting from accidental causes. It is always accompanied by a degree of inflammatory induration which extends more or less deeply on each side of the fissure; this is readily detected by the touch, the circumference of the orifice being uneven and lobulated. There may be one or more fissures, the intervening spaces being healthy or inflamed, excoriated, and in a state of erosion, or superficial ulceration. The fissures are often deep, and extend to a considerable distance in an upward direction.

The *discharge*, which is seldom so abundant as in the form of disease first described, has a decidedly purulent character, being alkalescent, of a yellow or greenish yellow color, sometimes brown or ichorous, and not unfrequently mixed with blood. This variety of ulceration is considered the most intractable—existing for years. Whitehead thinks it one of the most frequent **causes of** "habitual abortion." Whether this accident occurs habitually at the third, sixth, or any other period, is owing to the **extent** of the parts affected by the ulceration, and **Dr. A. K. Gardner**[*] adopts this view, and mentions many cases in proof. The *discharge* begins to be mixed with blood as soon as the expansion of the uterus extends itself as far as the diseased parts, the slightest irritation being sufficient to induce hæmorrhage. The careful introduction of the speculum, or digital examination, will cause bleeding. Hæmorrhage from fissured ulcer often gives rise to the **so-called** "menstruation during pregnancy," the habitual congestion during the usual menstrual period being sufficient to cause bleeding from the ulcerated **surface.** Regarding this phenomena, Dr. Whitehead thus writes: "The evidence now produced[†] appears sufficient to establish, as a general rule—to which I am not as yet acquainted with an exception—that the blood discharged in cases of alleged menstruation during pregnancy is furnished, not by the lining membrane of the uterus, nor by any healthy secreting surfaces—except sometimes the inferior part of the inner cervix—but by the lower extremity of the uterus, external to its cavity, or by the contiguous vaginal reflexion being in a state of *suppurative inflammation*. The fact is always demonstrable by the aid of the speculum."

[*] See cases in "Abortion and Sterility," p. 181.
[†] "Gardner **on Sterility.**"

"The *symptoms* indicating the presence of the disease in question, are—severe aching of the loins and sacrum, tenesmus, irritable bladder, with frequent desire to void urine, violent pain of the lower part of the abdomen, often confined to one side, at a point a little above the groin, corresponding to the situation of the round ligament, and following the course of the inguineo cutaneous and external pudic nerves."

Follicular Ulceration.—This form of disease of the cervix is said to be an inflammation and ulceration of the Nabothean follicles. It occasionally accompanies other morbid conditions of the cervix, but is most frequently met with in a distinct form. Whitehead doubts whether it is a primary affection. Upon *examination*, the part appears to be studded with a number of raised circular spots, having the dimensions of small peas, covered with a white crust, the surrounding surface being of a reddish hue. This white pellicle is easily removed by means of lint, exposing a surface of the same form and size, slightly elevated, and appearing as if composed of extremely minute granules. The parts are not painful to the touch. These spots are sometimes numerous, as many as fifteen or twenty being visible, but often only[*] one or two are seen. Owing probably to specific causes they sometimes become very prominent, and assume a warty appearance.

According to Bennet—who will not admit of any classification of inflammatory ulceration—this follicular ulceration is to be considered the same as any other arising from that cause; but he admits that "sometimes we find, in the vicinity of the os uteri, several small ulcerated patches, isolated one from the other, but near to it. These multiple ulcerations, which are rare, are

---

[*] See cases in "Abortion and Sterility," p. 181.

evidently formed, in the first instance, by *aphthæ, or ulcerated mucous* follicles."

I have under treatment, at the time of this writing, a case of follicular ulceration of the cervix, associated with hypertrophy of the uterus, and consequent retroversion. The patient had an abortion four months since, caused undoubtedly by these conditions. When this variety of ulceration occurs in cachectic habits, or in women advanced in life, these follicles are liable to inflame and suppurate in considerable numbers, leaving circular cavities, which give to the parts a worm-eaten appearance: or the suppurative action may extend, until several of the orifices coalesce, forming a deep, irregular shaped excavation, with callous, overhanging margins, constituting what some writers have described as the "corroding, or phagædenic ulcer." Induration and œdema of the cervix is a common concomitant of this aggravated form of follicular ulceration.

Phagædenic, Corroding, Cancerous Ulceration.—Under the above names have been described various forms of ulceration of the cervix, of a malignant character, characterized by a tendency to sloughing, with extension of ulceration to the body of the uterus, and a discharge, nearly always of a fœtid, corrosive, and otherwise unhealthy character. These forms of ulceration generally lead to organic changes, such as hypertrophy, induration, tumors, etc., which will be treated of hereafter.

Syphilitic Ulceration.—There can be no doubt that the syphilitic ulcer is a frequent cause of abortion. It was not until recently that its importance as such was suspected. In treating the abandoned classes in great cities, and even those in higher positions, the physician will do well to be on the look-out for this disease, when he is searching for the cause of abortion.

CAUSES OF ABORTION.

The *primary* venereal sore is very rarely seen attacking the lower part of the uterus. The virus is not likely to be carried up so high, except in occasional cases. It is chiefly *secondary* syphilis which affects the uterus. It may have its origin in three different ways, according to Whitehead.

(1) As an imperfectly cured primary affection, which was originally a genuine chancre in the external genitals.

(2) As a result of virulent inoculation upon the lower part of the uterus.

(3) As a consequence of secondary inoculation. The possibility of transference of venereal taint under this form, is still a matter of dispute.

The constitutional symptoms denoting the presence of secondary syphilis are, pallor of the countenance, languor, precarious appetite, loss of rest, hectic feverishness, lumbar and hypogastric pains, disordered secretions, and the appearance of the disorder in the offspring.

The local symptoms, enumerated in the order of their frequency, are, (1) *Endo-cervicitis*, or inflammation of the lining membrane of the cervix uteri, with inflammation, excoriation, or ulceration of the labia around the uterine orifice. This appearance was noticed in 19 out of 28 cases.

(2) A *mottled* or *patchy* appearance of the cervix, consisting of a number of dark red spots of irregular shape, surrounded by lighter colored portions: they sometimes appeared highly irritable and excoriated, but not aphthous; the whole cervix was generally enlarged and slightly indurated and in most cases there were evidences of endo-cervicitis.

(3) *Aphthæ* of the cervix, occurring in 8 out of 28

cases; the patches, which appeared perfectly white, of a circular or oblong shape, situated upon a dark red base, were easily detached by the aid of lint, and left a bright red surface of similar form and dimensions, having, in some instances, a very minutely granular aspect. These were associated with hypertrophy of the cervix and endo-cervicitis.

(4) *Warts* were witnessed in 3 out of 28 cases, twice on the cervix and once on the walls of the vagina. They were witnessed in many other cases.

The *syphilitic ulceration* of the cervix cannot be diagnosed from that of a non-venereal character, except by an actual knowledge of the presence of the specific poison in the system of the patient. There is nothing in the appearance of the ulcer or erosion which can show itself to be syphilitic. When we see *warts*, or the *mottled* appearance mentioned by Whitehead, we may justly have our suspicions.

Syphilis may act as a predisposing cause. (See that chapter.)

SECTION VII.

Induration of the Cervix Uteri.—This may be a primary or idiopathic affection. Whitehead met with severe cases of chronic induration during pregnancy, in only two of which treatment was effectual in averting abortion. All these persons had aborted; three of them, including the two just mentioned, he had attended in the preceding instances, in all of which the same form of disease at that time existed, and in the remaining four, it is probable, had long before existed. The sum of their abortions was 19; the average age of the patients 31 years.

This form of disease is usually met with as a chronic affection, the result of previous deep-seated inflammation, a species of prolonged metritis. The enlargement and induration may involve the whole neck, or may be confined to the anterior or posterior lip alone. In either case it may depend on inflammatory action, or upon some general cachexia. In the latter case the induration may be accompanied with hypertrophy due to fibrinous deposit, or infiltration of serum into the tissue of the cervix. The last named condition is termed œdema of the cervix. On *examination*, one or both labia of the cervix will present the appearance of a hard conical substance, sufficiently large, generally, to occupy the whole orifice of an average-sized glass speculum. If the anterior lip alone is affected, the os is situated behind, and high up, and is with difficulty brought into view. "When both labia are diseased they present a neutral appearance, very much resembling the top of the piece representing the bishop, in games of chess." Their opposing surfaces are often inflamed, excoriated or granular, and the fissures at their base ulcerated, and exuding pus.

This form of disease usually occasions abortion before the fifth month, although it may occur considerably later.

The *symptoms* of induration, hypertrophy, or œdema, are quite characteristic. A painful sense of constriction around the pelvis and hips is almost constantly present; a dull, fixed, aching pain of one or both ovarian regions, irritability of the bladder and rectum, pain in the region of the sciatic nerves, a highly excited state of the whole nervous system, lassitude, indigestion, etc. Habitual abortion may be the result of this diseased condition of the cervix.

Displacements of the Uterus.

The various malpositions of the womb may act as causes of abortion.

Anteversion, although of rare occurrence, may so irritate the bladder as to set up sufficient reflex action to result in expulsion of the ovum.

Prolapsus is a more frequent cause of abortion. This displacement generally occurs previous to pregnancy, as a result of congestion of the uterus, or induration, *with* hypertrophy of the cervix, and often from mere muscular atony. In either case the uterus will be found, on examination, low down in the vagina, the lips of the cervix resting on the perineal floor, and the body of the fundus within easy reach with the finger.

Until the end of the third month this condition is not generally provocative of any tendency to abortion, but about this time the uterus increases in size, and unless it rises above the promontory of the sacrum it becomes impacted in the pelvis. Any impediment to the proper expansion of the uterus will prove an exciting cause of abortion. In prolapsus and retroversion this impediment exists, unless the uterus rises out of the pelvic cavity.

Retroversion.—This is the most formidable displacement to which the gravid uterus is liable, and one which is more likely to cause abortion than any other. Retroversion during pregnancy was first fully described by William Hunter, and Denman. This affection occurs in its most decided form at about the third or fourth month, when the uterus is entirely within the pelvis, and when it is of such large size that any alteration from its natural position occasions great inconvenience to the neighboring organs.

Retroversion during pregnancy may be *sudden* in its occurrence; or it may have existed from the period of conception. In the former case it may be caused by unusual exertions, as lifting, jumping, reaching; or from such accidents as falling, jolting; or from diseases in which tenesmus occurs, as dysentery, etc.; also from distension of the bladder. In the latter case the retroversion had either existed previous to conception, or occurred during the first weeks of gestation. For a full description of the symptoms and causes of these displacements the reader is referred to my Monograph on that subject.*

Retroversion occurring during pregnancy is usually described as arising from excessive distension of the bladder, through neglect, reserve, or restraint on the part of the patient. It is believed that the full bladder, rising in the abdomen, drags the neck of the uterus upwards, while the distended organ presses the fundus uteri backward. The displaced uterus, by the pressure it exerts upon the neck of the bladder, in turn increases the distension of this viscus, and the retroverted uterus at length becomes fixed across the pelvis, the fundus lying in the hollow of the sacrum, and the os being tilted against the pubes, so as, in the worst cases, to render the evacuation of the rectum difficult, and that of the bladder impossible. Recent retroversion may undoubtedly be caused in this manner; but there are doubtless many cases in which retroversion has existed previous to pregnancy, and no inconvenience is felt from it until

* "Retroflexion and Retroversion of the Uterus."

such time as the uterus, from its size, presses upon the bladder and rectum to such a degree that the displacement is looked upon as *sudden*. In both cases abortion may occur from the irritation set up in the surrounding organs, and in the uterus itself, or from the utter impossibility of any further enlargement of the uterus.

The *symptoms* of retroversion, when the uterus is of such dimensions as to exert mechanical pressure upon the antero-posterior axis of the pelvis, are, in the first instance, partial or complete retention of the urine, pain in the pelvic region, and a sense of pressure on the rectum, giving rise to a constant desire for defecation, even when the bowels are empty. Should these symptoms pass unrelieved, the bladder becomes sometimes enormously distended, and even ruptured mechanically, or the coats inflame and ulcerate, allowing the urine to escape into the peritoneal cavity, and the patient sinks or dies of peritonitis. If the uterus cannot be replaced, and the water is occasionally and with difficulty drawn off, the bladder gradually enlarges and elongates, and its mucous membrane becomes diseased; muco-purulent, ammoniacal and bloody urine is passed, and the kidneys may become diseased by the effects of the backward pressure of the urine. The structures between the bladder and the uterus may become inflamed, and the patient be destroyed by irritative fever. In some cases all these mischiefs are arrested or modified by spontaneous abortion. In others the displacement continues to the fifth or sixth month, without destroying the patient, and it has been known to go on to the full term without causing a fatal result. These cases are however very rare, and the occurrence of abortion is less to be dreaded, than a continuance of the displacement.

CAUSES OF ABORTION. 87

Death of Embryo, from any cause, will result in an abortion. The presence of a foreign body in the uterus will, in the majority of cases, cause the uterus to exert its expulsive action. There are, however, many cases in which the fœtus and its envelopes have passed off in a fluid, disorganized state; and cases are on record even where the embryo after its death was undoubtedly absorbed. In another class of cases, the dead embryo itself, or the placenta, is left to become a mole or hydatid mass.

Coitus, when indulged in to excess, or even its moderate indulgence in women whose reproductive organs are irritable from disease, may be a cause of abortion. This may occur at any period of pregnancy, but especially at or about the usual menstrual periods. Coition causes abortion by exciting, from the mechanical irritation it induces, the reflex influences which preside over the uterus. If the os or cervix is diseased, the irritation excites the peristaltic action of the uterus; if not, it is probably the diastaltic action which is aroused.

Many cases of habitual abortion will be found to be due to this cause alone, and if abstinence from intercourse be strictly enforced, either wholly or at the menstrual periods, the usual abortions will be prevented.

As a rule, sexual intercourse should not be indulged in at all, or very rarely, during the period of pregnancy. That it is physiologically unnatural every physician will admit. The general aversion to the act manifested by women during that period, as also the fact that the animal creation rarely, if ever, cohabit during gestation, should teach us that it is improper. Whether sexual connection should be prohibited or not, will depend upon the judgment of the physician.

Instrumental Irritation.—The use of instruments, whether for *criminal* or *surgical* purposes, is a very frequent and powerful cause of abortion at any stage of pregnancy. *Criminal* abortion is rarely brought about by the administration of drugs; it is generally induced by some instrument which has for its object the puncturing or separation of the membranes, the death of the foetus, or the irritation of the os uteri, by which means the uterus is made to expel the product of conception prematurely. The same may be said when abortion or premature labor is induced by the physician, in a legitimate manner—*i. e.* to insure the safety of the mother or child, or both. Instruments have sometimes to be used during pregnancy, for purposes foreign to the accident above named, such as replacing a retroverted uterus, or in cases of extreme prolapsus, although such instances are rare. But there are cases on record where a retroverted uterus became so impacted that, other means failing, it was considered justifiable and necessary to use the uterine sound. In such cases it would seem that abortion would be the inevitable result; but I believe the operation might be performed in the early months, and the foetus and membranes remain intact. My opinion is based on several instances which have come under my own observation, where a sound was introduced into the uterine cavity, and turned several times round, with no other effect than slight pain and hæmorrhage—pregnancy not being interrupted, but terminating naturally.

Pessaries have been introduced for the purpose of supporting a prolapsed uterus during pregnancy. The presence of such a foreign body in the vagina, may, by exciting the reflex action, cause an expulsion of the ovum; yet pessaries have been worn for several months by pregnant women, without such a result.

As a general rule, the use of all instruments, introduced into the uterus or vagina, should be dispensed with, if possible, during utero-gestation, as well as all surgical operations on these organs, or on contiguous organs or tissues. Instances are not wanting, however, where women have borne the most severe operations, even on neighboring organs, without interfering with the normal advance of pregnancy.

Dr. Whitehead gives the treatment, by caustics, of several cases of ulceration of the os, during pregnancy, which treatment was successful, not only in curing the diseased condition, but of preventing an habitual abortion due to the ulceration.

No positive rule can be laid down as regards the use of such remedial measures, as the irritability of the uterus varies greatly in different individuals.

Ovarian.—The functional disorders which may prove exciting causes of abortion are *simple irritation*, from perverted physiological influences: *Congestion* and *Inflammation*.

During pregnancy, the ovaries should be in a dormant condition. Their functional activity is supposed to be suspended, because there is no use for the ovum, even if it were formed and extruded. In some instances the process of ovulation may go on during pregnancy, at the usual menstrual periods. We know the same process *does* often occur during lactation; but it is believed that in both cases it is an unnatural proceeding, and that it arises from some abnormal irritation of the ovaries. When this process occurs during pregnancy, the consequent afflux of blood, or nervous force, to the uterus, is often quite sufficient to induce abortion. We believe that this is the chief cause of that form of habitual abortion, which occurs within four weeks after conception, but about which so little has been written or known.

Congestion and *inflammation* of the ovaries may cause abortion by the severe reflex irritation which the diseased process originates. The *organic* diseases liable to cause abortion are, *ovarian tumors*, *dropsy* and *cystic enlargement*.

SECTION VIII.

MEDICINAL CAUSES OF ABORTION.

A list of the medicines which have been known to cause abortion at various periods of pregnancy; together with others which have been supposed to be detrimental to the life of the fœtus, has been given on a previous page of this work.

In order that we may arrive at some definite conclusion as to the *real* influence of the drugs named, it will be proper for us to consider them separately.

It is manifestly impossible for us to classify the medicinal causes, as we have classified other morbific influences, for in the present stage of our knowledge of the pathological action of medicines, we cannot decide with absolute certainty whether *secale* or *cimicifuga*, etc., cause abortion by their action on the spinal cord (centric), or upon the uterus directly (concentric). We cannot decide whether a medicine causes diastaltic or peristaltic action chiefly; nor can we say positively that *aloes* induces abortion by irritation of the rectum alone, or *terebinth* by its action upon the urinary organs.

But as all methods of classification must have a beginning, the following classes may be made, and their imperfections rectified at some future day, when greater pathogenetic or pathological progress shall have been made in our school.

Medicines which act as PREDISPONENT *causes.*— Belladonna, calcarea carb., carbo veg., china, chamomilla, chenopodium, ferrum, hyosciamus, lachesis, lycopodium, mercury, nux vomica, platinum, podophyllum, pulsatilla, plumbum, sepia, silicea, sulphur, zinc, and many others, which have the power of causing a dyscrasia in the system, or a cachectic state of the body.

Medicines which act as CENTRIC *causes.*—Secale corn., cimicifuga, caulophyllum, gelseminum, quinine, cannabis ind., nux vomica, ignatia, strychnia, are the principal medicinal agents at present known to the profession.

Medicines which act as CONCENTRIC *causes.*—All, or nearly all the medicines in the list may act in this manner; but upon various organs, and thence to the uterus by reflex action. Thus, *aloes, podophyllum, mercury,* and others are said to act on the *rectum; cantharis, cannabis sat., turpentine, sabina* (?) upon the urinary organs; and *apis mel.* upon the ovaries.

A medicine may act, however, both as a reflex and concentric cause of abortion at the same time.

Special analysis of the principal medicinal agents capable of causing abortion:

Apis mellifica.

This poison is stated to have produced " miscarriages in the second, third, and fourth months." Dr. Humphreys asserts, that he has in several instances known it to produce miscarriages when given for various maladies during pregnancy. This is said to be corroborated by other physicians. I have, however, known it to be administered with this intent by several physicians, but with no such result as above stated; and upon wide inquiry, I cannot obtain any reliable testimony relating to its oxytoxic powers. That abortion may occur while the patient was suffering from the poisonous effects of bee-stings, I do not doubt; but I beg leave to be sceptical as to the power of the 3rd and 6th dilutions, or even the 1st trituration, in causing that accident. If *apis* causes abortion when taken internally, it probably does so by its irritant action on the ovaries.

Actæa alba.

This is a near analogue of the *actæa racemosa* (*Cimicifuga*). It is called "white cohosh," in contradistinction to the latter, which is known as the "black cohosh." The country people allege that it causes abortion in the early months, and physicians have used it with success in suppressed menstruation.

Aloes.

The action of *aloes*, when given in large doses, in cases of amenorrhœa, is well known to every physician. It increases the menses, causes menorrhagia, congestion of the uterus, and *abortion*. This it is supposed to do, by causing an abnormal engorgement of the blood-vessels of the pelvic viscera. If it causes abortion, there is much coincident irritation of the rectum, tenesmus, hæmorrhoids, dysenteric symptoms, etc.

Asarum.

Asarum Europeum.—As early as the sixteenth century its abortion-power was spoken of by Hieronymus

Trajus. According to Berchtold, it is regarded in Bohemia as one of the most active and popular abortion-remedies. Dierbach places it with *sabina*, among the uterine remedies. It is used in England to cause miscarriage, and Hoffman cautions women who are pregnant, against its use. It seems to be an analogue of *pulsatilla;* an acrid, emetic, purgative medicine, irritating all the mucous surfaces.

Asarum Canadense.—This is a plant indigenous to the United States, known as "wild ginger," etc. Dr. Scudder says he was informed many years since that the root was used by the Indians as a "parturient, and also as an abortive." It was said to be used frequently by white women to prevent conception, and also to cause abortion. From these reports Scudder was induced to try it in amenorrhœa. In his hands it has acted well in restoring the menses.

Asclepias.

The *Asclepias Incarnata* was noticed in the *North American Journal of Homœopathy*, by Dr. Fowler, as being capable of causing abortion in small doses—viz., two or three drops twice a day; but the experiments of Dr. A. C. Jones, of Philadelphia, together with my own, entirely disprove this statement.

The *Asclepias Syriaca*, appears to possess a decided action upon the uterus, as its effects when given in large doses in a case of dropsy, reported in "New Homœopathic Provings," page 63, seem to prove.

Aletris farinosa.

This plant is said to have caused abortion with vertigo, headache, vomiting, and narcotic effects. Popular testimony ascribes to it the power of causing abortion in the early months; but we have no positive medical testimony to that effect. Professor Tully says very truly, that certain medicines are in successful use by the people for the induction of abortion, long before such uses become known and recognised by the medical profession.

Baptisia tinctoria.

This very depressing drug has been known to cause abortion. A strong decoction of the root, taken warm, at about the period of the menses, will produce such extreme relaxation as to result in miscarriage. Dr. Coe says, "It should not be used during the period of utero-gestation, as it is capable of producing abortion, for which purpose we have known it to be used by quacks and empirics. The danger to the general health is very great, when used in sufficient quantities to produce this result."

Borax.

This is a very ancient drug; known by the older physicians to be an "*ecbolic*," or producer of abortion, as well as capable of stimulating the inert uterus in cases of lingering labor. For the latter effect Kopp advises four to six grains every fifteen minutes, and says "it will arouse the expulsive action of the uterus, and terminate the delivery." Sunderlin testifies from his personal experience that it is useful in "difficult and irregular labors." Dr. Golding Bird says, "In women, this drug cannot be used with impunity, as it certainly exerts a stimulant action on the uterus, and I have seen it, in two instances, *produce abortion*." Premature labor has been caused by *borax*, 30 grains, three times a day, for 16 days. The labor was natural, and the child born alive.

"In doses of 10 grains, repeated three or four times a day, it has produced abortion, attended with pains all over the system, and excessive debility of the joints, which remained for several months, in a greater or less degree. On this account its administration to pregnant females is improper."*

Bovista.

This fungus is placed by Prof. Tully, in his class *ecbolica*, upon the authority of "a distinguished physician," (name not given) who "informed him that if collected before the interior became a powder, it was both a narcotic and ecbolic.

Hartlaub and Trinks give us a proving of *bovista*, which would seem to substantiate the above statement. It certainly has some specific action on the genital organs. We find some notable symptoms in the pathogenesis, among which are, "menses too *early* by eight or nine days, and more *profuse;* discharge of blood *between* the catamenia; bearing-down towards the sexual organs; *leucorrhœa*, etc."

It seems that this remedy has been too much neglected by our profession; or those who have tried it may have been disappointed in it. This may arise from the fungus from which the preparation was made having been collected when too late. We are told that the smoke or odor of the dry powder will act as an anæsthetic when inhaled; but the ecbolic property may exist solely in the green or ripe fungus, before it changes to a powder. Jahr's Pharmacopœia directs it to be col-

---

* King's Obstetrics, p. 714.

lected in the months of August and September. This is too indefinite. Experiments should be instituted to test its properties, and the most eligible method of preparation.

Cantharides.

This violent poison has been used for the criminal purpose of producing abortion, although its action on the uterus is uncertain and indirect. In cases where this effect has been produced, it was accompanied by intense inflammation of the urinary organs, swelling and heat of the generative organs, and violent general disturbance of the system. It has caused abortion when given to animals.

In the pathogenesis of Hartlaub and Trinks, it is said to "promote fecundity, expel moles, dead fœtuses and the placenta"—a rather sweeping declaration, and one quite unwarranted by the facts. A poison so uncertain in its action cannot be used with any advantage, in large or small doses, in such cases.

No more power as an "abortivant" can be claimed for *cantharides*, than for any of its analogues—*apis, cannabis, erigeron, copavia* and *turpentine*—except that it is the most powerful irritant. In fact *turpentine* seems more capable of *causing* abortion than any other. The *cannabis indica* is lately being used as a partus-acceleratur, and in persons predisposed to abortion, might excite that accident, as might any medicine, in large doses, which specifically affects the uterus.

Caulophyllum.

This medicine has a peculiarly specific action on muscular tissue generally, and that of the uterus in particular. While *secale* causes spasmodic *diastaltic* action, this drug seems to originate or call up the peristaltic motion of the uterus. It has been used since the settlement of this country, by the whites, and by the aborigines before them, as a partus-acceleratur. In very large quantities it has caused abortion, with strong expulsive efforts, but with no alarming or disagreeable symptoms of the general organism. But it requires such an inordinately large quantity of this agent to effect miscarriage, that it is not often used for that purpose. I have known one ounce of the tincture taken with no apparent effect, and a pint of strong decoction with like result. In fact, it often acts just the opposite of what is desired. I have known it resorted to by women, and even physicians, to hasten an impending abortion, with no other effect than to dissipate the pains, and other symptoms, and arresting the abnormal process. In these instances, one drachm doses of the tincture, and one grain doses of the alkaloid, were repeated every hour. The same may be said of many medicines of this class—*aletris far.*, *asclepias syr.*, *cimicifuga*,—they *prevent* abortion, even in massive doses. This should teach us, that although minute doses *may* suffice, we should not be afraid to resort to material doses, when sanctioned by clinical experience.

Cimicifuga racemosa.

That this powerful medicine will cause abortion at nearly all periods of pregnancy, the abundance of testimony, in and out of the profession, does not leave us in doubt. We do not mean by this to imply that the *black-cohosh* will *invariably* cause expulsion of the fœtus, for, like all other ecbolic remedies, it requires a constitution susceptible to its influence, in order to have its action developed. In the old American "Herbals" we are told that this root was used by the Indians to facilitate labor, and also by their women to produce miscarriage. Prof. Tully, one of the most extensive observers, has known it to produce abortion, and had great confidence in its power as a partus-acceleratur. He thinks all the *acteæ* possess the same power.

Since the publication of the work on *New Remedies*, I have received several letters from correspondents, good medical observers, who assert that they know the decoction of the roots, or the active principle (*macrotin*) will, in doses of four or six grains, cause abortion in some women, at any period of gestation. I have been cognizant of several instances in which women have induced miscarriage with the root, and also where physicians had administered the *macrotin* with that result.

The symptoms accompanying abortion caused by *cimicifuga*, are generally these: severe headache, as if the top of the head would be pushed off, or the eyes pressed out; stiffness of the whole body, with torpor, heaviness, and stiffness of the extremities, intermitting labor-like pains in the uterus, with pain in back and thighs, coldness and shivering, palpitation of the heart, neuralgic pains in limbs, contractions of the tendons, lowness of spirits.

If enormous doses are given, convulsions and serious injury to the system might ensue, as the remedy acts specifically on the spinal cord and brain; yet I cannot find any record of the drug having proved fatal.

The extensive experience which is being gained by our school, in the use of this remedy, proves that it is of great value in the treatment of many of the diseases of women, especially those occurring during the parturient state. It relieves the many pains and disagreeable sensations felt during gestation, prevents miscarriage, conducts accidental abortion safely, if inevitable; and it also controls false pains in the latter months, and facilitates labor, if tedious and unnatural.

Decodon verticillatus.

This plant, known to botanists as *lythrum verticillatum*, and by the common people as *swamp loosestrife*, or swamp-willow-herb, has the reputation of causing abortion in brute animals.

"If a great amount of testimony," says Tully,[*] "will decide anything in medicine, *decodon* is ecbolic for certain brute animals. This effect is said to be most frequently produced upon ewes, next upon cows, and sometimes upon mares. Any amount of testimony to this effect may be obtained from intelligent farmers, and even much from well educated physicians, who are either farmers themselves, or whose practice is in an agricultural region and almost wholly among farmers."

Tully thinks if this plant can cause abortion upon brute animals, it may produce the same effect upon women. Yet he is not aware that it has been verified. He instituted some experiments with it on the cat and dog, but with no definite result. He quotes Lindlay, who says "it is said to destroy the young of cattle heavy with calf."

King[†] says "it is said to cause abortion in mares and cows browsing it in winter, and may, perhaps, exert a medicinal influence on the female uterus" (uterus of women).

Culpepper (an old English writer) says it will "stay profuse courses in women."

[*] Materia Medica, Vol. 2, page 1369.   [†] Dispensatory.

The *decodon* may prove a valuable remedy when its powers are investigated and better known.

Gossipium Herbaceum.

I know of no agent used in medicine about which there is so much variance of opinion, as to its action on the uterus especially, as the cotton root. In my work on New Remedies* will be found the statements of Dr. Bouchelle, of New Orleans, and Dr. I. H. Shaw, of Tennessee, both of whom write enthusiastically of its powers as a uterine-motor stimulant, and mention as reliable, statements of its capabilities in causing abortion. Dr. King merely sums up the statements of other writers as to its alleged virtue, but gives no experience of his own. Dr. Coe's statements are of less value. Drs. Jones and Scudder believe it is emmenagogue, but "not abortional or parturiant in the slightest degree." Dr. Holcomb has no confidence in its pritudeal virtues. The writer has used it many hundred times and in all doses, but never saw it bring back the menses or cause abortion. Dr. Tully had no positive experience with it, but quotes Drs. Frost, Cabell, and Prof. Metlauer, as physicians who have tested it, and proved it to act powerfully on the uterus. Since the article in "New Provings" was written I have received a large amount of testimony, both for and against its asserted action on the uterus. Time only can clear up these conflicting statements. It may be that the usual preparations are inert, and only the *green fresh root should be used*.

Ilex Opaca.

The American Holly is mentioned by Prof. Tully as having the reputation of being a powerful "ecbolic."

---
* New Provings, page 217.

It is an evergreen tree, from 20 to 40 feet in height, and is found growing throughout the United States, from Maine to Louisiana, but is especially abundant in the district of Abbeville, S. C. Dr. Tully states that he was informed by several physicians of that region that it had a "high popular reputation as an ecbolic, it being considered capable of producing abortion or miscarriage at any stage of pregnancy." A strong infusion or decoction of the *leaves* is to be drank freely. Its use seems to be confined to the negroes. Dr. Tully could not ascertain that one educated physician had ever made a trial of it in a single case; yet he was told that no intelligent physician doubted its efficacy in that direction.

I cannot find in any work on materia medica or therapeutics any mention of the *ilex*, as a medicine specifically affecting the uterus. It has never been used in Homœopathic practice that I am aware of. It is to be hoped that some intelligent and investigating physician will test its pathogenetic powers, and add it to the list of our curative agents.

Mercurius.

Ever since the introduction of this deleterious drug, instances have not been wanting of its prejudicial effects upon the fœtus in utero. The older physicians observed that large doses of calomel administered to pregnant women, were frequently followed by abortion.

Stille[*] mentions that the inhabitants of districts where ores of *mercury* are smelted are seriously affected by its influence. "The annual mortality is more than 1 in 40. *Premature births* and *abortions are very com-*

---
[*] Materia Medica and Therapeutics.

mon." Dr. Lizi had an extensive observation of female operatives exposed to mercurial vapors. "Marriages among them were vastly more productive of *abortions*, stillbirths, and *feeble children* which seldom arrive at maturity, than among women engaged in other occupations." Dr. Colson, who wrote of the action of *mercury* on the uterus, says—"In not a few instances has it occasioned menorrhagœa or amennorrhœa, and in pregnant females *miscarriage*." More recent writers mention the fact that *calomel* frequently causes abortion. I was once informed by an Allopathic physician whose practice lay in a rather disreputable direction, that he knew of no surer producer of abortion than a massive dose of *calomel*—its hydro-cathartic effect was generally followed by expulsion of the fœtus. A woman once informed me that she usually arrested her pregnancies by a large dose of calomel at a menstrual period.

Podophyllum peltatum.

This active drastic cathartic has been known to cause abortion. I have been informed that the *podophyllin* is often administered for that purpose by certain disreputable Eclectic physicians. It probably acts in a similar manner to *aloes, calomel,* and other medicines which highly irritate the lower intestines, and by sympathy the organs of the pelvis. In large doses it is frequently and often successfully used in amenorrhœa, yet I cannot believe it has any such specific influence over the uterus as *caulophyllum* and some others. When threatened abortion is attended with piles, dysentery, prolapsus ani, bilious diarrhœas or extreme uterine prolapsus, this medicine will be indicated as a curative agent.

Quiniæ Sulphas.

In another portion of this volume will be found some mention of *quinine*. In the massive doses (30 to 40 grains) used in Allopathic practice, it has probably caused abortion, when given for the cure of malarious fevers. Dr. Petitjean affirms that he has so frequently seen abortion produced during the exhibition of *quinine* in intermittent fever, that he has ceased to prescribe it for pregnant women.* Dr. A. K. Gardner asserts that the administration of *quinine* to pregnant women is hazardous, and likely to cause premature labor.†

These statements are opposed by Rodriques and Henry, who attribute the abortion to the disorder of the general health, and the "mechanism of the paroxysm" during malarial fevers. Dr. Rich, of Georgia, attributes to it the *suspension* of contractions of the uterus threatening abortion under similar circumstances. By some physicians it has been alleged to render uterine contractions more active during labor, and as a means of overcoming rigidity of the os uteri.

Quinine is Homœopathic to many of the symptoms of abortion and its consequences, and will be found a useful remedy, especially in malarious districts.

Ruta graveolens.

This is a very old medicine, used by Hippocrates and other ancient physicians for a variety of diseases, among which was menorrhagia. Hippocrates, however, says it "excites the menstrual flow, and destroys the fœtus in utero." M. Hèlie, in 1838, published three cases of attempts to produce abortion by this plant, in one of

---

* *Lancet*, 4th, 1847.
† Tyler Smith's Lectures on Obstetrics, by Gardner.

which a decoction of the fresh sliced root, in the second a decoction of the leaves, and in the third, the expressed juice of the leaves was taken. The effects were, in one case, violent pain in the stomach, and vomiting, or rather efforts at vomiting, with rejection of a small quantity of blood. In all of the cases the nervous system was prominently deranged; there was great prostration, with confusion of the mind; cloudiness of the vision with feebleness of pulse; cold extremities, and twitching of the limbs. All of the females, who were in the fourth or fifth month of pregnancy, *aborted* and recovered. Hahnemann records the effects of *ruta* thus: "Metrorrhagia? Miscarriage? Sterility?" It would seem that he doubted its power of causing abortion: a doubt which other observers do not share. Stille says it is emmenagogue, and "not only like *ergot* acts upon the gravid uterus, but it stimulates the unimpregnated organ also." M. Bean recommends it, associated with *savin*, in *uterine hæmorrhage* after abortion, and *menorrhagia* from general debility.

Sabina.

The Juniperus Sabina is a native evergreen shrub of Southern Europe, and is cultivated as an exotic in gardens in this country. The Juniperus *procumbens*, a species having very similar, if not identical, properties, is found growing on the shores of our north-western lakes. I have met with it in the interior of Michigan, and also on the sand-dunes which cover the southern shore of Lake Michigan. It has been found on the shores of Lake Huron. It throws out its branches very close to the ground, spreading in large circles, generally quite horizontal, but often growing upward at a slight angle of perhaps 10 or 20 degrees. This indigenous variety should be proven, in order to test its comparative effects.

Dioscorides was the first to describe the qualities of *savin*. Among other things, he says, "An infusion of them in wine causes bloody urine, and applied as a fomentation to the belly of pregnant women, they produce *abortion*." Galen says it "Excites the menses more powerfully than any other agent, provokes bloody urine, *destroys the life of the fœtus, and causes its expulsion*." "But," says Stille, "it was most celebrated for its emmenagogue properties, which were habitually invoked for the criminal purpose of destroying the product of conception." This belief of the power of *savin* to cause abortion has been perpetuated to this day, and it is now frequently resorted to for that criminal purpose, notwithstanding the almost certainly fatal consequences resulting from its administration in doses sufficient to cause the destruction and expulsion of the fœtus. The annals of medical jurisprudence abound in cases of fatal poisoning by *savin* taken to produce miscarriage. It is much more employed than is commonly imagined for criminal purposes, but fortunately in such doses as fail of their purpose, and only produce instead vomiting and purging. It generally produces inflammation of the intestines and urinary organs, and sometimes congestion of the brain. A case is reported by Mohrenheim of a pregnant female who took an infusion of *savin* to produce abortion. It caused incessant vomiting, and some days afterwards excruciating pains, abortion, flooding, and death. Rupture of the gallbladder was found on examination of the body, and an effusion of bile into the abdominal cavity, with peritonitis. Many other cases might be quoted. Hahnemann records several of an interesting character, by which he was led to recommend its use in cases of abortion. In most cases enteritis-mucosa, and peritonitis were detected after death. Tully says, "The active principle of *sabina* has exactly the same composition as *oil of turpentine*, if it is not identically the same in all respects." It is a curious fact, apparently confirmatory of this statement, that the external application of both *savin* and *turpentine* to the abdomen of pregnant women, will often cause abortion. Both agents seem to have a similar effect on the intestines, urinary organs, skin, and general organism.

Sabina, in poisonous doses, acts most powerfully upon plethoric women, whose menses are habitually profuse. In such constitutions it would be more apt to provoke an abortion, than in persons of spare habit, even if they menstruated profusely.

Secale cornutum.

Although *ergot* has been used as a uterine motor stimulant since the year 1096, yet its *modus operandi*, as well as its general and local action, seems far from being understood.

It seems quite settled, however, that it will cause *intermittent* contraction of the uterus at full term. That it is capable of causing abortion seems to be generally doubted. Says Stille: "If we examine the influence of *ergot* upon the gravid uterus in the earlier stages of pregnancy, and before quickening, when it is most likely to be resorted to with criminal intentions, we shall find that in proportion as we *recede from the period at which the spontaneous action of the womb begins, ergot fails to exhibit its peculiar influence.*" M. Dargan, in his report to the Academy of Medicine, says: "We do not believe that, independently of labor, of direct manipulation, or of some other external influence, that *ergot*, of itself, can excite uterine contractions during the first half of pregnancy." Stille remarks that this is probably the opinion of physicians who have had the best opportunities of studying the subject. It is, however, to be recollected—he says—that the distension of the uterus at this period is a normal condition, and it does not follow that an equal or even a less degree of enlargement should not have a different result when it depends upon morbid causes. This is the case in menorrhagia. It was so in a singular accident reported by Dr. Taylor, of New York. A female who was not pregnant had some leeches applied to the neck of the uterus to relieve engorgement of that organ. One of them found its way into the cavity of the uterus, where it occasioned bearing-down pain and a bloody discharge. *Ergot* was administered, and a clot expelled holding in its centre a dead leech. According to Dr. Ker, an *enlarged* and prolapsed uterus contracted under the use of *ergot* so that it could be placed *in situ*. If we examine the annals of criminal abortion, we find that enormous doses of *ergot* have been taken, sometimes with fatal effect upon the mother, and often to the production of serious disease, without causing the expulsion or death of the fœtus. In other cases the *life* of the fœtus was destroyed, but it was not expelled. Two very interesting cases of poisoning by *ergot* are published in the "Transactions of the New York State Homeopathic Society, vol. ii.," in one of which 160 grains of *ergot* were taken, causing a serious illness, but without "any pains or signs of labor," and pregnancy went on uninterrupted. In the other case, reported by Dr. Hill, half an ounce of *ergot* was taken to produce miscarriage, by a woman who was evidently not pregnant, but affected with uterine hæmorrhage due to irritation of the uterus. In this there were "pains of an expulsive character in the uterus" the first day, and continued "pain in the vulva" nearly every day until she died. I have known it to cause expulsive pains in passive menorrhagia.

Effects of ergot upon gravid animals.—It is important that we should study this point. Stille says, "The influence of *ergot* upon the gravid uterus in animals is not uniform. In some cases it seems to have been purely negative, in others to have destroyed the product of conception without producing its expulsion, and in still others—and these are the most numerous—to have caused abortion. Bonjean gave *ergot* to a female *guinea pig* during the early stages of gestation; abortion was not occasioned, nor, indeed, any symptom whatever. The experiments of Wright were to this effect. He mixed *ergot* with the food of a pregnant rabbit; no tendency to abortion was excited, and in due time six healthy young ones were born. The animal, still kept upon the same food, was again impregnated. She looked drowsy and moped, the fur grew erect and rough, gestation was *protracted* beyond its usual term, and three young ones were born, two of which were dead, and the third survived but a few hours. The same experimenter, after many trials of *ergot* upon pregnant bitches, concluded that it did not act as a parturifacient in them, although it sometimes appeared to injure or destroy the product of conception.

On the other hand, according to Diez, it produced abortion in bitches and sows, without harm to the mother or the young, when the dose was moderate, but large doses destroyed both, and excited inflammation of the womb.

Dr. Oslen gave *ergot* to a *sow*, a *cow*, and a *cat*, before the completion of pregnancy, and in each case produced abortion. In 1841 an epidemic of abortion among cows occurred in France, which was traced to the ergotted state of the rye and other graminæ in the district.

The most of the above cases of abortion in animals undoubtedly occurred during the last half of pregnancy. The same effect would have been produced in women under similar circumstances. After the study, observation, and experience the writer has bestowed upon *ergot*, his belief is that it is capable of causing abortion in animals and women, in the early months of gestation, in those constitutions which are susceptible to its action. I do not believe, however, that it excites contractions and expulsive pains *until it destroys the life of the fœtus*. When this occurs the enlargement of the uterus becomes an *abnormal* condition, because it contains a foreign body, which is a source of irritation of itself. In this condition of the uterus the *ergot* will act, and cause expulsive pains and contraction. In the last months of pregnancy it may induce premature labor, but the child is generally still-born, so that the exceptions to the above rule are very few.

It may be said of *ergot* that it is a dangerous and uncertain medicine, when used to cause abortion, and should never be resorted to for that purpose.

Sanguinaria canadensis.

This medicine is one possessing active and varied properties, and is capable of exercising a strong effect upon the uterus—whether direct or not cannot yet be satisfactorily decided.

Rafinesque cautions against its use during pregnancy, as it acts very powerfully on the uterus, and causes *abortion*, and recommends its employment in amenorrhœa. Tully says it is sometimes emmenagogue, and has been known to produce uterine hœmorrhage.

Many eclectic practitioners assert that they have used it successfully in amenorrhœa and infrequent menstruation. They state that a sure sign of its remedial action is the presence of "pains in the small of the back, extending down the thighs." These would probably be the symptoms in abortion caused by blood-root: there would also be other important symptoms—as vertigo, pain in the head, narcosis, vomiting, burning pains in the stomach, etc.*

Terebinth.

The oil of turpentine is one of the most active and powerful of that group of medicines which affect the urino-genital organs. The apparent starting-point of its action seems to be in the urinary apparatus, but it produces profound aberrations in the nervous and vascular systems. Persons exposed to the vapors of turpentine experience its general effects, such as nausea, vertigo, impaired vision, pain in the back and loins, strangury, bloody urine, insomnia, malaise and an eczematous eruption. Women, in addition, are often affected with *menorrhagia* or dysmenorrhœa. [Eminent Allopathic authorities claim for it great curative powers—which it really has—in all hæmorrhages, especially those from the *uterus!*] Stille† says it arrests hæmorrhage after parturition by exciting contractions of the uterus, as it is known to do, as well as by its styptic

---

* See New Homœopathic Provings: Art. Sanguinaria.
† Materia Medica: Art. Tereb.

(homœopathic) powers. *Turpentine* when taken internally has been known to cause ABORTION and *premature labor*, accompanied or not by the above-mentioned local and general symptoms. A patient who applied to me for the cure of sterility, informed me that she had caused abortion several times in the early months upon her own person, by simply rubbing the hypogastric region with the *spirits* of turpentine. Several years had elapsed since she had been pregnant; her menses were too profuse and frequent. By this it would seem to cause both abortion and sterility.

Tanacetum vulgaris.

This common plant has been perhaps more frequently resorted to for the purpose of inducing criminal abortion, than almost any other agent. The *oil* of tansy is the preparation usually selected for this purpose. The many cases of death arising from the use of this powerful poison do not seem to deter the vicious or unfortunate from its use.

In overdoses, the *oil* causes unconsciousness, flushed cheeks, dilated pupils; hurried, stertorous respiration, strong spasms, full and frequent pulse, repeated convulsions; then failing pulse and death. In a case of attempted abortion, by a decoction of the herb, there occurred delirium, slow and laborious respiration, contracted pupils, dusky countenance, fixed features, and cool skin. Subsequently the muscles of deglutition and all the voluntary muscles became paralyzed, and death with gradual retardation of the heart's action, took place in 26 hours after the poison was taken. It would appear that the operation of the tincture or decoction of *tansy* was not the same as its essential *oil*. Both preparations have undoubtedly been used successfully as abortivants, without fatal or serious results; but its administration in large quantities is generally dangerous. In the February number of the *North American Journal of Homœopathy* (1865) will be found a *resumé* of all that is known concerning the pathogenetic and curative effects of *tanacetum*.

Ustilago madis.

This *ustilago* is a parasitic mushroom, which occurs on maize (Indian corn) as *ergot* does on rye. In a cow-house where cows were fed on Indian corn infested with this parasite, eleven of their number aborted in eight days. After their food was changed, none of the others aborted. The better to be convinced of the poisonous nature of these mushrooms, the author, after having dried and pulverized them, administered six drachms to two bitch dogs with young, which soon caused them to abort.*

Lindlay says:—"Its action on the uterus is as powerful as the *ergot* of rye, and perhaps more." According to Roulin—"Its use (long protracted, of course) is attended with shedding of the hair, both of man and beast, and sometimes even of the teeth. Mules fed on it lose their hoofs, and *fowls lay eggs without any shells.*"

Tully in his mention of this fungus adds—"It is doubtless by its abortifacient power that it causes the eggs of fowls to be extruded before there has been time for a shell to be formed. By what power does it cause the shedding of the hair of man and brute animals and the casting off of the hoofs of mules long fed upon it?" It would seem to be capable of great curative powers.

* Anal. Med. Vetr. Belge. and Rep. de Ph.

PART III.

GENERATION:

SYMPTOMS, DIAGNOSIS, PATHOLOGY, MECHANISM, AND PROGNOSIS OF ABORTION.

SECTION I.

GENERATION.

The Physiology of Generation is so intimately connected with the subject of Abortion, that it would be improper to omit some mention of that process, when conducted normally.

Generation, in its broadest sense, is that function of the female generative organs which dates from the successful impregnation of the ovum to the period of its birth. It includes the several processes of contact of the ovum with the fertilizing semen of the male, its passage through the Fallopian tube into the uterus, certain textural changes in the uterus in advance, and in consequence of its reception and developement, and its final passage and parturition.

I propose to treat of *Generation* as including three distinct stages, namely,

(1) *From the arrival of the impregnated ovum in the uterine cavity until the placental attachments are perfected, or the period of quickening.*

The *ovule* has its origin in the ovary, and when it has attained its full maturity, the vesicle in which it is enclosed becomes the seat of an excitation, which finally results in a rupture of the walls of the vesicle, and the extrusion of the ovule. This maturation, and escape of the ovule, generally occurs at or about the menstrual periods. After its escape, the ovum engages in the Fallopian tube, the enlarged extremity of which has been applied to the ovary. It has been supposed that the above processes occur from the stimulus of coition, or erotic excitement, as well as from the excitation consequent on the menstrual crisis; but it is doubtful if the former uniformly causes such results.

The ovum passes through the Fallopian tube, but the time necessary for this passage is not with any certainty known. In the human subject, says Cazeaux, no one case has ever proved its existence in the womb prior to the twelfth day. But this cannot be accepted as any decision of the question.

When the ovum enters the cavity of the uterus, if it has been fecundated by the spermatic fluid in its passage, it attaches itself to some portion of the hypertrophied mucous membrane, generally near the fundus, at which portion the *placenta* is afterwards attached.

It is here necessary to make some mention of the *decidua*. It is now argued that at each menstrual period the uterine mucous membrane is exfoliated, thrown off, and a new one formed in its place. It is this membrane, greatly hypertrophied, that we find in membraneous dysmenorrhœa. It is now well established that the decidua is nothing more than the hypertrophied mucous membrane. Whilst the evolution of the ovarian vesicle is going on in the ovary, the vascularity of the uterine mucous membrane is greatly increased, and the highly congested vessels are discoverable beneath the epithelium. This state of turgescence, however diminishes during the last days of the menstrual epoch, and disappears almost entirely sometime after the catamenia has ceased. But if the ovule, before leaving the ovarian vesicle, or during its passage through the tube, receive the vivifying influence of the spermatic fluid, the fecundation will maintain and increase the abnormal excitement of the genital organs, and then, instead of subsiding, the uterine mucous membrane becomes still more turgescent, of a deeper violet color, and the folds and wrinkles increase so as to more than line the cavity of the uterus. It is in one of these folds that the ovum is enclosed, after the lapse of a period as yet unascertained, and this fold of hypertrophied mucous membrane which remains in the uterus, when united, forms the *decidua reflexa* of authors. The ovum is also enveloped in its own proper membranes—the *amnion* and *chorion*.

From the date of conception and lodgment of the ovum in the uterine cavity, until about the fourteenth week (three and a half months), its nutrition is carried on by means of imbibition, or absorption through the membranes that surround it. *Up to this period the placenta is not attached to the uterus*, and the connection between the mother and child is established by means of the allantois. The fact that *before the fourteenth week of gestation the placenta is not attached to the uterus* should be kept in mind, as it will have considerable bearing on the pathology and treatment of abortion. This period is also that known to writers as the period of "quickening," or a time when the fœtus is connected with the maternal circulation through the placental vessels. In a medico-legal point of view, this is an important date, as by some jurists it is considered a date after which the induction of abortion is considered a criminal offence, unless some point of medical expediency demanded it.

(2) *The second stage extends from the date of the attachment of the placenta to the uterus, until the period when the fœtus is capable of a separate existence.*

The beginning and end of this period, however, can not be said to have any fixed limits. It is supposed by some that the period of "quickening" corresponds with the attachment of the placenta. To a certain extent this may be true, as both generally occur about the sixteenth week.

The age at which a fœtus is viable may be said to be about the end of the seventh month. Cases occur in which children born at an earlier time than this survive, but these are rare exceptions. (Dr. J. W. Francis reports a fœtus born with membranes intact in the twenty-third week of pregnancy that lived to maturity.) Even at the seventh month they are kept alive with difficulty. The state of the heart as regards development, the feebleness with which the fœtus sucks before this time, the ready failure of animal heat, and the inability to bear the movements necessary to nursing and cleansing, render it almost impossible to rear the fœtus born at an earlier period.

(3) *The third stage of utero-gestation reaches from the viability of the fœtus until the end of pregnancy.*

This, however, like the period of viability, is a variable period. According to those authorities who consider the last day of the last menstruation the proper date to reckon from, the termination of pregnancy varies from the thirty-seventh to the forty-third week. Dr. Reid, whose elaborate calculations appeared in the *Lancet*, gives the terminations of 500 pregnancies, which ranged from the thirty-seventh to the forty-fifth week. Dr. Reid also calculated from "a single coitus" in 43 cases, all of them resting upon testimony as credible as can be obtained in such cases. These ranged from 260 to

300 days, giving the average duration of gestation at about 275 days. Nine months is supposed to be the average duration of human pregnancy, but the time undoubtedly varies from eight and a half to ten months.

Dimensions and weight of the fœtus at the different periods of uterine life.

A treatise on Abortion would be incomplete if it did not contain some practical information on this subject. In a purely medical, and especially a medico-legal point of view, this is of manifest importance.

At the time when the embryo first begins to be distinct, that is, about the *third week*, it is oblong, swollen in the middle, obtuse at one extremity, though drawn to a blunt point at the other, and straight, or nearly so, being somewhat curved forward. It is therefore vermiform in shape, of a grayish white color, semi-opaque, almost without consistence, and gelatinous, varying from *two to four lines in length, and weighing one or two grains*. At this period the only trace of the head is a small tubercle separated from the rest of the body by a notch, but no rudiments of extremities are observed, nor is there a cord at first.

At the *fifth week* the embryo becomes more consistent; the head is large in proportion to the body; the rudimentary eyes are indicated by two black spots turned toward the sides; it is nearly *two-thirds of an inch long, and weighs about fifteen grains.*

At the *sixth week*, the bronchial fissures disappear, leaving only a slight cicatrix, and its size and weight are somewhat increased.

At the *seventh week*, the first centres of ossification appear. The intestine still extends for a considerable distance along the umbilical cord. At this time the embryo is *nearly an inch in length*.

At *two months*, the forearm and hand can be distinguished, but it is not supplied with fingers. The cord has not yet assumed a spiral arrangement; it is four or five lines in length, and is inserted near the lowest point of the abdomen. It is very difficult at this period to distinguish between the sexes, owing to the extreme length of the clitoris. The embryo is from *one and a half to two inches long, and weighs from three to five drachms.*

At *ten weeks*, the embryo is from *one and a half to two and a half inches in length*, and weighs an ounce or an ounce and a half. The cord is longer than the embryo, and begins to assume the spiral arrangement, but its base always contains a portion of intestine. The fingers are distinct, but not the toes. At the end of the *third month*, the embryo weighs three to four ounces, and measures from *five to six inches*. The cord contains no intestine; the nails begin to appear, the sex is distinct, the eyeball is seen through the lids, the forehead and nose are clearly traced, and the lips well marked and not turned outward.

At the *fourth month*, the embryo takes the name of *fœtus*. The body is *six or eight inches in length, and weighs from seven to eight ounces*. The face still remains but little developed; the eyes, nostrils, and mouth are closed; the skin has a rosy color, and begins to be covered with down, and the muscles now produce sensible motion. "A fœtus born at this period," says Cazeaux,* "might live for several hours. Whilst I was Interne at the Hotel Dieu, I received one that had

---

* Cazeaux's Midwifery, page 208.

scarcely reached the fourth month. It lived, however, from half-past seven to half-past eleven o'clock."

At *five months*, the *length of body is eight to ten inches, and it weighs from eight to eleven ounces.* The skin is more consistent; the pupils cannot be distinguished.

At *six months*, the hair is longer and thicker, the nails are solid, but the eyes are still closed. The *length* is *eleven to twelve and a half inches*, and the *weight* about one pound (avoir.)

At *seven months*, the eyelids are partly open, and the testicles begin to descend into the scrotum. The fœtus acquires a *length* of *twelve and a half to fourteen inches.*

At *eight months*, it is only *sixteen or eighteen inches* long, and yet *weighs four to five pounds*, because the fœtus seems, at this period, to grow rather in thickness than in length. The scrotum generally contains one testicle, usually that on the left side. The skin is very red, and covered with long down. The lower jaw is now as long as the upper one.

Finally, *at term*, the fœtus is about *nineteen to twenty-three inches long*, and weighs from six to seven pounds. Cazeaux thinks the weight and length of children at birth have been wonderfully exaggerated, in which he is probably correct.

SECTION II.

SYMPTOMS OF ABORTION.

The symptoms of an abortion may be divided into *three* classes, namely:

1. *The* PREMONITORY.
2. *The* ACTUAL.
3. *The* SUBSEQUENT.

Each of these classes, or divisions, constitutes a different stage of the miscarriage.

The PREMONITORY symptoms constitute the first stage of the abortivant process, namely—*the stage of irritation, or that condition of the uterus which exists up to the time of the rupture or separation of the membranes or placenta.*

The ACTUAL *symptoms constitutes the time which intervenes from the separation of the membranes, etc., until the expulsion of the fœtus and placenta.*

The SUBSEQUENT *symptoms are those which follow and mark the conditions which are known as the sequelæ of abortion.*

It is important that the above divisions be borne in mind, as it leads to a methodical study of the subject, and has an important bearing upon the treatment.

The *symptoms* of the premonitory stage may be said to include all those symptoms which belong to the *causes* heretofore mentioned; but more strictly considered, are those which occur for a few days or weeks previous to the commencement of the second stage.

The premonitions of an abortion may be present for a long time, or only for a few hours. This depends upon the nature of the causes; for if the cause be an ulcer on the cervix, the symptoms may appear at every monthly period, and may not result in actual expulsion of the fœtus if proper remedial means are used; or, should the cause be a fall, concussion, or instrumental, the premonitory symptoms may be few, or entirely wanting.

The *prodromes* of an abortion generally appear in the following order:

1. *Pain.*—This may consist merely of an aching in the back (sacrum) or hypogastrium, and extending down the thighs; or it may be acute, and described as griping, lancinating, darting or stitching; but whatever the character of the pain may be, it is generally confined to the above localities.

2. *Sensations*, which are not pains, but consist of a feeling of *weight, pressing-down*, or *soreness*, in the hypogastrium and in the pelvis. These sensations may co-exist with the *pain*, or may be present without the pains above mentioned.

3. *General symptoms.*— There is almost always a general uneasiness, nervousness, languor, and depression of spirits, accompanied or not with some acceleration of pulse, flushed face, and cold extremities. At the same time the symptoms which are usually attendant on pregnancy remain.

4. *Hæmorrhage*, generally a slight discharge, which may continue days, and even weeks, before the second stage begins.

But so soon as the membranes are broken, by accident, or punctured by artificial means; or the membranes or placenta become separated from the walls of the uterus; or the fœtus dies from any cause; then we have a train of symptoms in addition to the prodromic, namely:

1. *Chills.*—My observation and experience goes to show that the chill, or rigor, whether preceded or not by pains and sensations, is the most reliable symptom of a *rupture* or *separation* of the membranes. The fœtus may die, and remain for some time in the uterus before it excites irritation sufficient to arouse the expulsive action of that organ. Meanwhile the placenta may be said to *live*, and even perform its functions, as in certain cases when the blighted or dead ovum is changed to a *mole*.

In other cases we may be made aware of the death of the embryo by certain symptoms generally obscure; but until the woman has a chill, or chilly sensations, we may consider separation of the membranes as not having occurred. This *chill* greatly varies in intensity and duration. Sometimes it consists of a vague sensation of internal chilliness or coldness, and may last for days, and may be mistaken for the first stage of an influenza or a fibrile attack; at other times and in other patients, it may assume the form of *rigors*, in which the woman will shiver and shake, as during a paroxysm of ague. I have seen instances, even, where the attack could have been easily mistaken for a "congestive chill," so excessive was the prostration and the general coldness.

There is a class of cases in which *rigors* may appear *unattended* by any sensations of coldness. These are called, by old nurses, "nervous chills." The woman will shiver, her teeth will "chatter," and she will appear to suffer from great chilliness; but will tell you she "is not cold." This form of "rigor" is usually met with in cases of parturition at full time, and is supposed to indicate a *relaxation of the circular muscles of the cervix*, or of any sphincter muscle. I am inclined to view this as a correct explanation. In one instance in which I noticed this rigor without coldness, the embryo was expelled in the unbroken membranes, together with the placenta. In this case, the time which intervened between the separation of the membranes, and the expulsion of the whole, was too short for the appearance of a chill.

2. *Pain.*—As in the prodromic stage, the pains may be aching, cutting or griping, but they are generally attended with another kind known as "labor-pains." A woman may feel a pressing-down sensation in the pelvis during the premonitory stage, but that sensation is quite different from the one under consideration. A labor-pain is a bearing-down sensation *accompanied* with pain: this pain regularly intermits, or remits, which is rarely the case with the premonitory sensation alluded to. Some authors claim that it is possible to arrest the progress of an abortion in the second stage. I think it hardly possible, as I have never known such an arrest to take place after the occurrence of intermittent, labor like pain, coming on *after a chill*. From the above, it will be noticed, that I consider the second stage of abortion to be marked by two prominent symptoms, namely: the *chill*, and the *labor-pain*. There is another prominent symptom of the second stage. I allude to the painful sensation of *soreness*, tenderness, or sensitiveness of the hypogastric region, and sometimes the whole abdomen. This is often so severe as to lead to the belief of the presence of peritoneal inflammation, which, however, rarely occurs in such cases: or more correctly to the occurrence of metritis, which is often

present to a certain extent. This sensation is not always due to inflammation or congestion; but is oftener *neuralgic* in its nature, having its seat in the abdominal muscles, or even in the uterus, which is only a hollow muscle. But as this symptom sometimes attends the prodromes, and as the abortion is often prevented after its occurrence, it cannot be considered as belonging exclusively to the second stage.

Pains may exist in other portions of the body, as the back, thighs, hypochondria and head. The pain in the head is particularly to be noted: it is often intense, and affects principally the top of the head, and the eyes, and is described as a painful pressure from within outwards (a pain quite similar to that caused by *cimicifuga* or *macrotin*). This pain in the head, as well as the nausea and vomiting which sometimes occur, is due to reflex irritation. Sensations of numbness, lameness, and cramps of the upper and lower extremities, sometimes appear, and are due to the same cause.

A common symptom of the second stage is a sensation as if the back (sacrum) was dislocated, or as some patients express it, "as if they *had no back* in one place," alluding to the sacral region. It is during this stage that we sometimes find considerable febrile excitement, occurring generally subsequent to the chills. This fever may come on in irregular paroxysms, and be followed by perspiration, and so nearly simulate certain forms of ague, that the careless practitioner is often misled as to its real significance. Together with the appearance of the above symptoms, we generally notice a disappearance of the usual symptoms of pregnancy, or those which the patient has usually been troubled with, if she has borne children. The morning nausea and vomiting subsides; the "longings" cease, and the enlarged breasts become soft and flabby.

Hæmorrhage.—This is almost invariably attendant on the second stage. There are, however, some few exceptions to the rule, as in the case mentioned above, of the expulsion of the unbroken membranes. In this instance absolutely no hæmorrhage occurred, as I was informed by the nurse, who stated that only the slightest stain of blood appeared upon the napkins used. *Hæmorrhage rarely occurs to any great extent before the expulsion of the fœtus:* it is from this occurrence, up to the final expulsion of the placenta, that flowing is most to be feared. I believe that no instance of fatal, or even dangerous hæmorrhage can be cited as having occurred previous to the expulsion of the fœtus. It is rarely the case that the placenta is expelled at the same time with the fœtus, or shortly thereafter; while the contrary obtains in delivery at full time. I have often thought that the expulsion of the fœtus ought to mark an intermediate stage, between the second and the third. Many hours or days, and even weeks, may elapse between the expulsion of the fœtus and the placenta. Meanwhile the patient generally loses considerable blood, which may flow uninteruptedly, or the flooding may occur at longer or shorter intervals. The chills, which marked the onset of this stage, may occur frequently, at irregular intervals, or if the patient resides in a locality where malaria abounds, the paroxysms may occur regularly, and assume all the characteristics of an idiopathic intermittent. It is not strange, perhaps, that intelligent physicians have treated such cases as pure agues, and overlooked the intra-uterine cause. *Anti-periodics*, or *quinine*, while they will break the *regular* recurrence of the paroxysms, will not cause them to subside entirely, nor will any medicine arrest them until it first causes the expulsion of the retained placenta.

The placenta is frequently retained in the uterus until putrefaction takes place, in which case it is not *expelled*, in the usual acceptation of the term, but passes away in a dissolved or disintegrated state. In such cases the discharges from the uterus have a peculiar and persistent fœtor, unlike anything else—a fœtor which the physician who has ever inhaled it will never forget. The odor of the discharge from uterine cancer bears some relation to it; but there is a marked difference, which the experienced practitioner can detect. It is peculiar to both discharges, that it is almost impossible for several days to eradicate entirely the disagreeable fœtor from the hand which has been used in making a necessary examination. The best preventive of such a disagreeable contamination is to annoint the hand thoroughly with fresh lard, previous to the examination. The lard absorbs the odor, and retains it, in the same manner as it absorbs the delicate and costly perfumes of flowers, which are placed between layers of purified lard, in order to preserve volatile odors, and transmit them to pure alcohol.

The decaying placenta may be weeks in passing away, or it may be expelled by the irritated uterus before this process is completed. In this case the mass expelled has a spongy, worm-eaten appearance, and exhales an intolerable effluvia.

In some instances the placenta is neither expelled or discharged in putrefactive solution, but in some manner keeps up a connection with the uterus, and continues to enlarge, becoming in the end a hydatid or molar mass.

There are even cases where the after-birth has remained in the uterus a long time, having no connection with that organ, remaining about the size it had attained when the fœtus was expelled, yet undergoing no change of a putrefactive character, and finally being thrown off with or without hæmorrhage. One such case came under my own observation. A woman aborted in the third menstrual period from conception. The fœtus was expelled, considerable hæmorrhage followed, and it was supposed the placenta had been thrown off. *Ergot* was given to arrest the flooding; nothing further occurred for a month, when violent hæmorrhage occurred, and again two weeks after. At this time I was consulted. *Caulophyllin*, 1–10th, in doses of two grains every two hours, arrested the flooding, and under its use for twenty-four hours, the unchanged placenta was expelled. No fœtus was present at the time nor afterwards. It is supposed that in these instances, putrefaction is prevented by the closure of the cervix so tightly as to be hermetically sealed.

We may here inquire why the placenta is so often persistently retained. But one explanation is usually given, namely, that the uterus, before the fourth month, is quite undeveloped, and its muscular structure incapable of originating or maintaining contractile or expulsive action. That this is often the cause of the non-expulsion of the placenta is not to be disputed, but that it is *always* the cause of its retention is certainly not the case. Those who have attended many cases of abortion, if they are at all observant, must have noticed how frequently the uterus is, in such instances, more or less *retroverted*. It has seemed to me that this dislocation occurs in two-fifths of all cases of abortion before the fourth month. In such a malposition the cervix is flexed, and even bent at nearly a right angle, sufficiently so as to nearly or entirely close the canal of the cervix; and no amount of effort can expel the placenta until the uterus is placed in proper position. Anteversion of the uterus will have the same effect, but as this occurs more rarely, it is not of as much importance. I have met with but three cases of this latter form of dislocation occurring during the progress of an abortion. The uterine sound was here efficient in effecting the change of position, and dislodging the retained placenta.

A woman who has aborted may suppose she has got rid of the whole contents of the uterus, and her physician may be of the same opinion, unless he has had considerable experience and is a close observer. There may be a slight discharge, bloody or not, or there may be none at all. This condition of uterine quiescence may continue for days, and even weeks, when suddenly after walking, stooping, lifting, or some unusual exertion, uterine pains, with or without hæmorrhage, may set in, and a retained placenta be thrown off.

If the patient only has pains while lying *on her face*, we may generally consider that retroversion exists. This I have observed in several instances, and upon examination, I found that the uterus changed to a natural position when the woman lay upon her face, and dropped into the hollow of the sacrum when she turned upon her back.

After the expulsion or discharge of the placenta and membranes, if no coagula are present, the chills and fever generally cease, but there is, however, a kind of *irritative fever*, which may occur before or after this period. It is probably due to the presence of a substance undergoing putrefaction in the uterine cavity, and the absorption of morbid matter into the circulation. This fever, although zymotic in its character, is not, like puerperal fever, due to any specific contagion, nor is it propagated from one woman to another, *i. e.*, my experience does not lead me to suppose that any such contagion is to be apprehended. I have often gone direct from the room of a patient from whom I had just extracted a putrid retained placenta, to attend a woman in labor, using no more than usual means of cleanliness, yet I never had a case of puerperal fever in my practice under such circumstances.

This putrefactive, or irritative fever, is usually attended with all the symptoms of a typhoid, namely: the heat of the skin, quick, irritable pulse, dry tongue, stupor, or coma-vigil, and even diarrhœa. But it is a notable fact, that if at any period of the fever the uterus gets rid of its morbid contents, the unfavorable symptoms disappear with surprising rapidity, showing that the condition of the blood is not due to any fermenting poison working in that fluid, but from the *absorption* of a poison, only.

The *complications* which may ensue during this fever, are inflammation of the uterus, phlebitis, ovaritis, pelvic cellulitis, and occasionally cystitis; but as these more usually occur as sequelæ of abortion, they will be considered under that head.

The *discharges* from the uterus before all the morbid material is thrown off, are often irritating and excoriating in the extreme, causing in their passage outward superficial erosion, and even ulceration of the os uteri, vagina, and vulva. I have now described the symptoms of abortion, in various degrees of severity, from the first premonitus to the entire expulsion of all the products of conception. The *subsequent* symptoms remain to be considered.

The *sequelæ* of abortion are many and important, comparing in gravity with those of an unnatural labor.

It is the opinion of some authors that serious consequences are oftener the result of abortion than of premature labor, but this statement is hardly warranted by the facts. If abortion is properly treated, the sequelæ are very few and unimportant. It is only when this accident is improperly treated, or left to the unaided powers of life, that serious results occur. The most common sequelæ are metritis, ovaritis (acute and chronic), induration and ulceration of cervix and os uteri, leucorrhœa, prolapsus, retroversion, chronic metrorrhagia, and anæmia. To these may be added the occasional occurrence of mastitis, peritonitis, and puerperal mania.

The symptoms of the above diseases are supposed to be familiar to every practical physician, and are to be found in every work on diseases of women. We will therefore omit to enumerate them in this place. In relation, however, to mastitis, it may be said that the mammæ rarely become engorged and filled with milk before the third month; but after that time it is no uncommon occurrence to have all the symptoms appear which usually accompany the secretion of milk at full time, and even of the occurrence of mammary abscess.

SECTION III.

DIAGNOSIS OF ABORTION.

From the numerous and characteristic symptoms just given, the *diagnosis* of abortion ought to be very easy; but unfortunately, these signs are not very clearly marked until the abortion is inevitable, and consequently when it is a matter of indifference to the patient whether the physician makes out a clear diagnosis or not. It is therefore during the premonitory stage, that we should endeavor to recognize the true nature of the disease, for then only can our art succeed in arresting its progress.

The diagnosis of abortion involves the solution of three questions:

(1) Is the woman pregnant? If she is,

(2) Are the symptoms those of a commencing abortion, or do they arise from other diseases?

(3) Is the abortion inevitable?

Is the woman pregnant? This first question is quite readily solved after the fourth month of gestation, though before that period it is almost always unanswerable. All physicians of experience are aware of the difficulties which involve it. A woman in good health may cease to menstruate for several months; she will show nearly all the natural signs of pregnancy. At the third or fourth month she may have signs of uterine congestion or irritation, lasting for several days, followed by a slight flow of blood. Is it a return of the interrupted menses or an approaching abortion? The physician should try to satisfy himself, if possible, of the actual existence of pregnancy (see "Conduct of Physician"), but if this cannot be done, he must rely upon the symptoms present.

If *hæmorrhage* occurs, we must distinguish it from those rare cases of "menstruation during pregnancy," so-called. This is supposed to be an exudation of blood from an ulcerated os or diseased cervical canal, or from the lower segment of the uterus, not occupied by the deciduous membrane. To distinguish it from this abnormal form of menstruation, we must ascertain if the symptom has occurred every month since the symptoms of pregnancy set in, and also the duration of each previous hæmorrhage. If such has occurred, and this subsides like them, it is plain it cannot be an impending abortion, unless such hæmorrhage proceed from *placenta previa*, as is sometimes the case.

Little or nothing can be inferred from the form and size of the *clot*, whether it has proceeded from an unimpregnated womb or not; but all clots should be examined, by carefully picking them in pieces under clean water, and if the abortion has occurred after the third week, the embryo may be discovered.

Cazeaux gives certain rules, laid down by Holl, how to distinguish a clot in the cervix uteri from the head of the fœtus, but our space will not permit their insertion.

(1) *Metritis* may occur idiopathically during pregnancy, when it is pretty sure to cause abortion. If it occur from medicinal or mechanical causes, and for the purpose of causing criminal abortion, the consequences are very grave.

To distinguish an impending abortion from inflammation of the uterus (unimpregnated), we must consider the character of the pain, which in the latter is *not* intermitting, nor is there any hæmorrhage. A metritis may be followed by a fœtid grumous discharge, but it lacks the peculiar fœtor of a decaying decidua or placenta. Finally, the history of the case, and the absence of the usual symptoms of pregnancy.

(2) *Peritonitis* may occur without causing abortion. The first symptoms of this disease, however, may be mistaken for symptoms of abortion, particularly the abdominal tenderness, the chills, and the tympanites. But the absence of intermitting uterine pain, the condition of the os and cervix, and the absence of hæmorrhage, will enable us to form a correct diagnosis. As in metritis, we may have both a peritonitis and an abortion existing at the same time.

(3) *Dysmenorrhœa* has many symptoms which very closely resemble those of abortion, so nearly, indeed, that it is almost impossible to form a satisfactory diagnosis between that disease and an abortion before the tenth week of gestation.

The shreds and skinny substances discharged in membraneous dysmenorrhœa, may not contain any discoverable fœtus or placenta. But if, as I believe to be often the case, abortion occurs in the third or fourth week after conception, the embryo is so small as to elude a very close search. In fact the deciduous membranes expelled in dysmenorrhœa and early abortion are said by recent investigators to be identical, and sometimes their expulsion is attended with all the symptoms of abortion. It may be said, therefore, that there are many cases which come under the care of the physician, where it is impossible to give a decided opinion one way or the other.

(4) *Dysentery*. The pains which accompany dysen-

tery, and are located in the hypogastrium and sacrum, extending in some cases down the thighs, the tenesmus, and desire to bear down with the abdominal muscles, so nearly simulate the symptoms of an abortion, that the latter has often been prescribed for, as dysentery, especially if a diarrhœa has been present *with* the abortion,—a not uncommon occurrence. But no careful and observing practitioner will ever be guilty of such carelessness.

When pregnancy exists, may the symptoms be attributed to simple congestion, or should they be regarded as the first tokens of a threatened abortion? Although it is very difficult to decide this question within the first three or four months, or at the beginning of the accident, its solution is happily of little importance, as regards the treatment, the measures indicated for simple congestion being equally applicable as preventives of of miscarriage.

"When symptoms, which in all appearance were due to simple congestion, have yielded to proper treatment, the physician is often required to answer a question whose rigorous solution is always impossible—namely: the abdominal and lumbar pain being allayed, and all the other alarming symptoms removed, is the patient therefore out of danger of miscarriage? In the majority of cases we can tell nothing about it, for it is impossible to know whether the congestion has been arrested in time to prevent a rupture of blood-vessels, and an effusion between the placenta and uterus, or whether the separation of the placenta is extensive enough to have destroyed the fœtus immediately : even supposing the child to be still living, we cannot ascertain the degree of separation of the placenta, nor foresee the effect which a partial destruction of its maternal attachment may have upon the fœtus. Very frequently, indeed, the latter, by being cut off from a considerable part of its means of respiration, is placed in the condition of an adult, whose lungs are in great measure destroyed, and whose respiration and nutrition being insufficient, gradually wastes away. As the child often does not perish until after the lapse of eight days, two weeks, and frequently not until the next menstrual period, this, too, without the appearance of any new symptoms to explain the unlooked for death, the physician, therefore, cannot be too reserved in his diagnosis, as regards the possible consequence of such accidents."*

But if the abortion has really begun, can we hope to arrest the symptoms? Severe pains, their constant direction from the umbilicus towards the occyx; the previous duration of the discharge, and the amount of blood already lost; softening and dilatation of almost the entire neck, and even of the internal orifice, and projection of the membrane during contraction, all indicate a very unfavorable prognosis. It is said by some authorities, that these symptoms should not destroy all hope, but I have never known abortion arrested after it has reached this stage. It is even stated that rupture of the membranes, and discharge of the amniotic fluid, does not render abortion inevitable. But this assertion is simply absurd; for such a condition not resulting in death, and expulsion of the embryo or fœtus, is impossible. In the cases alluded to by Desameaux, there was certainly a mistake in reference to the true origin of the *water* lost by the patient.

Hydrorrhœa, resulting in discharge of water from the uterus, has undoubtedly been mistaken for rupture of the ovum. Cazeaux relates a case where the occurrence took place at three and a half and four and a half months; the pregnancy terminated naturally.

* Cazeaux's Midwifery.

Hæmorrhage may occur without an abortion being inevitable, for it may arise from an ulcerated os, a diseased cervix, or even a slight separation of the placenta. The amount of discharge is more important than its duration. A slight hæmorrhage may continue for several days or weeks, since it may originate in the rupture of a few vessels. I have known it to last six weeks or two months without compromising the pregnancy. But if a large amount of blood is lost in a very short time, the placenta must be separated to a considerable extent, and abortion must necessarily ensue.

Abortion is really inevitable only when the fœtus has ceased to live, when the membranes have been broken, or when the separation of the placenta, and the rupture of the utero-placental vessels, are so extensive that the remaining utero-placental attachments are unequal to the support of the fœtal respiration. It is impossible to ascertain in the early months whether the fœtus is living or dead. The sudden cessation of the vomitings, salivation, swelling of the breasts, and other sympathetic functional disorders of pregnancy, are pretty sure proof of the death of the fœtus. The continuance of these symptoms, even after the occurrence of leucorrhœa and other disturbances, is certainly favorable.

There is a particular form of the neck of the womb, which Cazeaux says is only met with when abortion has taken place. "When the patient has been for a short time only pregnant, we know that it is always easy to distinguish the neck of the uterus from the body; in the great majority of cases we may even feel the angle which separates them. Now when the contractions have lasted for a certain length of time, they have gradually dilated the internal orifice, the cavity of the neck has become confounded with that of the body, and when the finger in the vagina is passed over the entire lower segment of the uterus, the neck can no longer be distinguished from it; a well-defined limit between them is no more to be detected, and all that belongs to the neck of the womb has the shape of a pear, the larger part being continuous with the body of that organ, and the lower extremity corresponding with the external orifice. Whenever I have met with this condition of things, abortion has taken place."*

After the fourth month of pregnancy the diagnosis is much more certain, there is greater hæmorrhage, and dilation of the os is more easily detected, and the death of the fœtus can be verified in a positive manner. The following are the signs of this occurrence. (*a*) The abdomen diminishes instead of increases in volume; (*b*) the breasts shrink away; (*c*) the woman has dragging sensations about her loins, an unusual weight in the hypogastrium, as of an inert body which falls toward the side on which she lies from the mere law of gravity; (*d*) the movements of the infant cease to be perceptible. (*e*) Lastly, the most valuable evidence is that furnished by auscultation, for an impossibility of hearing the sounds of the fœtal heart after the fifth month is an almost certain sign of the child's death; indeed it is the only sign, for all the others may be absent, and yet the fœtus be living. Unfortunately, the pulsations of the heart are not generally perceptible before the fourth month of pregnancy.

We may, in most cases, be able to diagnose the existence of an abortion during its progress, and directly after the accident; but we are often called upon to treat

---

* Cazeaux does not mean to imply that that the fœtus and membranes have been expelled, but that their expulsion is inevitable.

the sequelæ of abortion, and, were we not closely observant of, and acquainted with, the symptoms occurring from retention of decidual debris, or a placental mass, we might treat the patient for some other malady.

Retention in Utero of the Ovum, Placenta, Decidual Membrane, or parts of either.

It is well known to the profession, that the whole or portions of the product of conception may be retained in the womb, after the vitality of the ovum has ceased. In such cases, a putrid discharge generally occurs, which is sometimes attended with danger to life, and which generally disappears after a longer or shorter time.

We are often called to cases presenting the following array of symptoms: The woman is much prostrated, anæmic, with sometimes an icterode hue of the skin; she is very languid, hysterical and depressed in spirits; she may be confined to her bed, or tries to be up and attend to her domestic duties: in the former case there is continual hæmorrhage, or a constant sanious discharge, having an abominably fœtid odor; or in the latter case the hæmorrhage occurs occasionally, at irregular intervals, coming on suddenly, with or without pain. We sometimes find the symptoms very severe: with the prostration there will be loss of appetite, tumid or tender abdomen, frequent small and sharp pulse; hot and parched state of the skin of the hands and feet, hectic fever and night sweats. The discharge is extremely fœtid, and there are frequent hæmorrhages, brought on by the slightest exertion.

After examining a case of this character, the physician will be likely to pronounce that an abortion has occurred, and the ovum, placenta, membranes, one or all, have been retained, and are undergoing putrefaction. But this would not always be a correct *diagnosis*. There are other causes of the above symptoms, all or in part, namely: retention of the lochia, of leucorrhœal or purulent discharges, detached polypi, cancer of vagina or uterus, extra-uterine pregnancy, disintegrating fibrous tumor, abscess of the genital organs or pelvis, and thrombus or hœmatocele of the same parts. These causes do not *invariably* produce all the above symptoms, or *any* of them, but they all *frequently* do so. The same may be said of tents, pessaries, or any substance introduced from without, which will decompose, or lead to retention of matters that readily undergo putrefaction. The art of distinguishing the retention of the ovum and its appendages from the above, is one which can only be learnt by close study and much experience. The scope of this work will not permit a further reference or an extensive differential diagnosis. The physician must judge from the history of the case. He should first satisfy himself whether pregnancy previously existed; other matters should follow after.

There are other consequences of imperfect abortion which differ from the above symptoms, in that an early ovum, or a part of it, may be retained for months, without causing any discharge having a noticeable fœtor, or indeed any discharge at all. These cases are the most difficult of diagnosis; there may be no special symptom to distinguish the illness from an ordinary uterine ailment; even an examination by the touch or speculum will fail to help us. A case is reported by Dr. Duncan where the ovum was retained *seven* months, with the "absolute absence of any fœtor." She suffered during that time from "weight in uterine region, slight bearing down, occasional irritability of the bladder, irregular action of the bowels, occasional disorder of the stomach, and even sickness, *brownish* leucorrhœa, bright bloody discharges, often profuse, never absent above a week. At seven months a sponge tent was inserted into the cervix, and the next day the ovum was found in the vagina. It was a placental mass an inch and a half broad, and above half an inch thick. On opening the bag of membranes, a few drops of dirty brownish fluid escaped; no remains of an embryo or cord was discoverable, and the ovum perhaps never contained any, being addle from the beginning."*

SECTION IV.

PATHOLOGY AND MECHANISM OF ABORTION.

"This is a subject," says Dr. Duncan, "which is little understood." He probably alludes to the changes which take place in the uterus, embryo, placenta, decidual membranes, etc., during the *process of abortion*, and not to the *causes* of the abortion. The latter are pretty well understood; not completely, however, for there are many predisponent and exciting causes which are probably yet unknown.

The pathological changes which go on in the uterus and its contents, during an abortion, varies with the stages of utero-gestation. Abortion, when it occurs very early in pregnancy (before the twentieth day, during which time M. Guillemot calls it *ovular* abor-

---

* Edinburgh Medical Journal, January 1863, p. 589.

tion), is generally owing to certain obstacles which prevent the permanent attachment of the ovum to the uterine walls. In cases where conception has occurred just *before* a menstrual period, the motor act of expulsion is probably limited to the Fallopian tubes, the ovum being carried out of the uterus with the menstrual discharge.

When conception has occurred just *after* a period, the ovum may be dislodged by some mechanical action, or motor irritation, affecting the uterus, in which case it would fall into or near the cervix, and after undergoing disintegration, pass off with a leucorrhœal discharge.

After twenty days, and until the third month, or fourteenth week (embryonic abortion), after the decidual membrane is fully developed, and before the placenta has formed its uterine attachments, the great vascularity of the uterine mucous membranes renders the effusion of blood between the decidua and uterine walls an easy occurrence. This extravasation may arise from simple congestion, from rupture of a vessel, or from separation of the membranes by instrumental means; but from whatever cause it arises, the blood collects and spreads in all directions, separating the decidua from its connections, and causing contractions in the uterus, which generally end in the expulsion of its contents.

In cases occurring in this stage, the canal of the cervix and the os uteri have to be dilated before the ovum can pass, and this process of dilatation occupies a considerable time, and frequently causes much suffering. The most favorable way in which an early abortion can occur, is when the detachment of the entire ovum takes place before the act of expulsion occurs. The perfect ovum is then expelled at once, and the uterus contracts without much hæmorrhage. In other cases the membranes are ruptured, and the small foetus comes out alone or enveloped in the amnion, or the membranes may be discharged piecemeal, leaving the ovum to escape afterward. As a general rule, the membranes remain after the expulsion of the embryo, and the earlier the abortion, the longer the placenta or membranes have a tendency to remain. This is probably owing to the extended adhesion of the ovum to the internal superficies of the uterus, and the feeble power of the uterus to contract on its contents. Sometimes the membranes of an early ovum will remain for weeks, but in such circumstances there is not the same tendency to decomposition and its dangers, as there is in a case of placenta after the sixth month.

I have observed, in reading some reports of cases of abortion before the *third* month, the remark that "the placenta was adherent." Such a condition cannot exist prior to that date, for the reason mentioned above, that until that time the placenta does not form its attachments to the uterine walls.

Although the contractile power of the uterus at this date is comparatively feeble, yet it is sometimes quite notable. This contractile power has been greatly underrated. Dr. Simpson has seen the uterus contract, when unimpregnated, upon the uterine sound. The virgin uterus contracts violently during dysmenorrhœa, and with labor-like pains expels the abnormal decidua ("false membranes"). I have known the uterus to contract forcibly, during an abortion, before three months, and the pains were very like those of labor. This contractile power of the uterus, even at an early date, may be taken advantage of when we are striving to cause expulsion of the placenta.

The appearance of the masses discharged in early abortion, is described with excellent minuteness by Dr. Meyer.* He says:

"When these masses do not prove to be mere coagula, they present the following appearances: "In form they resemble internal coatings of the uterus, the fundus and cervix being quite discernable. The external appearance of the mass is that of a coagulum of blood with a more or less smooth surface; and this it is found to be on cutting into it, until we arrive at about its middle, when we come upon a cavity having smooth walls, more or less collapsed. Upon nearer examination, this cavity is found to be lined with two membranes, the chorion and amnion. An affixed funis is always found, and near its attachment, the umbilical vesicle, and frequently the *ductus omphalo*-mesariacus. The free extremity of the funis has all the appearance of being torn. These various appearances fix the age of the fœtus at about *two months;* but no fœtus is to be found, or even the fragments of one."

The conclusion usually come to, that no fœtus has been present, was regarded by Dr. Meyer as inadmissible, and he therefore instituted a more exact investigation into these cases. He accordingly found in all of them a rent, extending through the membranes, usually at the place which corresponded to the orifice of the uterus, and this rent led into a canal of greater or less length, amidst the external coagulum. So constantly is the funis directed towards this rent, that in one case where the placenta was implanted more towards the orifice of the uterus, and a large rent had occurred at the fundus of the ovum, the funis passed directly up-

* Henle's Zeitschrift, Band, X. p. 283.

wards. It is evident, then, that the fœtus escapes through the rupture of the membranes, and the following seems to be the proximate cause of its doing so. Abortions of this kind are complicated with considerable hæmorrhage, and the blood effused between the walls of the uterus and the ovum, whether in the fluid state or as a coagulum, when acted upon by the uterine contraction, compresses and bursts the ovum. The membranes collapse, and the funis becomes fixed in the position it assumes on the exit of the fœtus through the rupture. So small an object as the fœtus becoming mixed with the coagula is easily overlooked.

Dr. Duncan writes, in his paper "On some of the Results of Imperfect Deliverance in Abortion:"*

"In abortion it sometimes happens that the entire double layer of decidua is discharged with the ovum; in this case the abortion may be truly called complete. It also sometimes happens that the ovum alone is discharged, unaccompanied by any decidual structures; and in such cases the incomplete abortion is followed, after a few hours, or even a day or two, by the expulsion of the remaining decidual masses. Occasionally no such decidual masses are discharged as masses, and yet recovery is undisturbed; and in cases of this kind the persistent decidual membrane must either disintegrate rapidly, and come away imperceptibly in the discharges, or, maintaining its uterine connections, the membranes may slowly exfoliate, and atrophy in like manner as it does after ordinary menstruation. But it also, though rarely, happens that the decidual masses are retained for many weeks undecomposed (perhaps adherent to the uterus), and then become separated, putrify, and cause fœtid discharges, until they are expelled."

In abortions occurring from the time of the maternal attachment of the placenta until the viability of the

* Braithwaite, Part 47, page 234.

fœtus (the sixth month), the *mechanism* resembles more closely the pains and motor action of a natural parturition, and the tendency, as regards the expulsion of the ovum, is to imitate labor at full time. The *pathological* changes are the same as in abortion, occurring during the first stage of gestation, with this additional feature, that the extravasation may occur between the placenta and uterine parietes, as a result of intense congestion (placental apoplexy), mechanical separation by instruments, or strong contractions of the womb. When this occurs the membranes are also separated by the out-pouring blood, and the uterus thereby irritated until its motor actions are aroused; the cervix uteri is slowly dilated, the membranes ruptured (if intact before), and the fœtus expelled, to be followed at a longer or shorter interval by the membranes and placenta. As gestation advances the cervix uteri becomes developed, the difficulty of passing through the cervix becomes diminished, while that of passing the pelvis is increased. When the ovum is small, the contractions of the uterus are chiefly or solely concerned in its expulsion; but when it is large enough to distend the vagina, the abdominal and respiratory efforts are called into play. From the sixth to the ninth month the pathology and mechanism are nearly the same, but more closely simulate natural labor.

SECTION V.

PROGNOSIS OF ABORTION.

Unless the abortion has been caused by violent means; or the use of instruments in the hands of the the patient, or an unskillful or reckless physician; or from serious organic disease, which has previously prostrated the vital powers—the prognosis of an abortion *may generally be considered as favorable.*

An important element constituting a favorable prognosis, is the rational and scientific treatment of this accident. An abortion may begin in a favorable manner; the uterus may do all that is demanded of it for the purpose of expelling the embryo; but the treatment adopted may, by deranging the functions of other organs, or arresting the natural efforts of the uterus, render the result of the case decidedly unfavorable. On the other hand, a case may commence with the most alarming symptoms; the uterus may fail to put forth any proper efforts; yet the skillful physician, by using the medicinal and instrumental means appropriate to the case, may conduct it to a safe and rapid termination.

The period at which an abortion occurs influences the prognosis. Some authorities, among them Desamoreaux, assert that it is more serious for the patient in the last stages of gestation. This is not always the case. I am inclined to agree with Cazeaux, that "It hardly constitutes an indisposition during the first and even second month, but in the third or fourth, the expulsion of the fœtus demands a certain dilatation of the os uteri, and tolerably energetic contractions, for the neck and body of the uterus have not yet undergone the modifications necessary to such an effort, and the delivery of the after-birth often presents difficulties less frequently met with at a more advanced stage of gestation; whence I conclude that an abortion is then more grave and painful to the patient, as also more dangerous, than in the fifth or sixth month."

Tyler Smith states, that "In abortion the danger from hæmorrhage is before the expulsion of the ovum: in labor at full term, it occurs after delivery." The reasons for this opinion have been given when treating of the Pathology of Abortion.

As I have before stated, dangers rarely occur in abortion before the sixth month, after the placenta is expelled or has been removed.

A favorable prognosis may be given in all cases when the fœtus and placenta has been expelled with but little hæmorrhage at or after the occurrence, or when the placenta has been removed before severe hæmorrhage has occurred, or even if the secundines cannot be removed but pass off in putrefactive solution unattended with a low grade of fever.

It is unfavorable when, in cases of criminal abortion, such violence has been used as to cause serious inflammation of the uterus, etc., or injuries to that organ or contiguous structures. The uterus has been lacerated or perforated by sharp instruments, and death has resulted from internal hæmorrhage or peritonitis; sharp probes have been forced through the bladder, or into the cul-de-sac between the rectum and vagina, causing serious and sometime fatal results. Caustic or acid fluids have been injected into the womb and caused death by metritis; or if the substances injected passed through the Fallopian tubes, fatal peritonitis has ensued. Finally, the internal administration of such poisons as *ergot*, *sabina*, and *turpentine*, have caused such intense uterine inflammation, and constitutional disturbance, as to destroy life.

The instances above enumerated are usually those which result from criminal ignorance or recklessness, and the injurious causes have their origin in the first stages of abortion. There is, however, another class of cases, which might perhaps be termed *natural* abortions, as they are due to some disorder of the body. In all these cases an unfavorable prognosis can only be feared in the event of two morbid conditions having obtained, viz:

(1) The retention of the placenta with flooding.
(2) Its putrefactive absorption.

In cases of the retention of the placenta, if it be not removed, severe, protracted, and dangerous flooding may ensue; death may then occur of sheer exhaustion from loss of blood unless the placenta is removed.

In other cases of retention, it is not the flooding only that is to be dreaded, but the absorption of fluids in a state of putrefaction. In quite a large experience, however, I have never lost a patient from either cause; in fact, I have never lost a patient from abortion or any of its consequences. Cases, however, do occur, of death from flooding, or irritative fever, as also from some of the sequelæ of this accident.

Two cases will illustrate how a little medicinal or instrumental interference may change the most unfavorable case into one of only moderate danger, and insure final recovery. I was once called to see a woman said to be dying. Three allopathic physicians had attended her for six weeks, and diagnosed "inflammation of the womb with gangrene." She appeared nearly *in articulo mortis;* pulse scarcely perceptible, face hippocratic, skin cold and covered with a clammy sweat. The discharge per vaginum was most intolerably offensive. I got a hasty history of the case, but sufficient to satisfy me that an abortion had occurred about eight weeks previously. *China*, in thirty-drop doses, was alternated half hourly with the same quantity of the wine of *ergot*. Brandy and food were given freely. In six hours a horribly offensive placental mass was expelled. The patient made a rapid recovery. The other case was a nearly similar one, except that death was imminent from profuse flooding, which the tampon or medicines had been powerless to arrest. The blunt hook was introduced, and the placenta removed; hæmorrhage ceased immediately, and the woman had a favorable convalescence.

We have considered the *immediate* prognosis, and it will be noticed that we do not coincide with the old proposition which has been advocated since the time of Hippocrates, viz., that the prognosis is more grave than that of labor at full time. But the *remote* consequences are undoubtedly more disastrous in the former case. Thus the acute diseases which attack lying-in women are more frequent after labor, whilst the chronic disorders of the genital organs which appear in advanced age, are more common in females who have often aborted than in those who have been delivered at term. Again, it is highly important to notice the unfavorable influence which one abortion has on subsequent pregnancies, for whenever a woman has had a miscarriage she is more predisposed than others to a similar accident, and hence great precautions should always be taken to prevent it.

"The prognosis," says Cazeaux, "as regards the *fœtus*, is always fatal." This author, however, limits abortion to the period preceding the time of viability fixed by law, namely, the end of the sixth month. He admits that cases are reported of children born prior to this period which have lived; but these examples, he says, even if they were authentic, are too rare to invalidate his general proposition.

PART IV.

TREATMENT OF ABORTION.

SECTION I.

TREATMENT OF ABORTION.

The *treatment* of abortion may be divided into

1. PREVENTIVE.
2. REMEDIAL.
    a. *Mechanical.*
    b. *Medicinal.*
3. POST-PARTUM.
    a. *Postural.*
    b. *Dietetic.*
    c. *Medicinal.*

This excellent division is the one adopted by my colleague, Dr. Ludlam, in his Lectures on Obstetrics, and I have appropriated it as the most methodical and scientific which has come under my observation.

I. PREVENTIVE.

The preventive treatment of abortion consists manifestly in the remedial measures adopted for the removal of those diseases which have been enumerated as being the causes which have a tendency to result in the death of the ovum; the separation of the membranes and the expulsion of all the products of conception.

(1.) *Constitutional or Predisponent.*

Plethora.—If this condition is caused by an excess of nutritive material taken into and assimilated by the organism, the proper treatment would seem to be the adoption by the patient of that diet which would most effectually cut off the supply of tissue-making material. It is the opinion of the best physiologists of the present day that the ingestion of starchy or saccharine matter directly tends to cause corpulence. It matters not whether starch or sugar be taken into the stomach as such, or whether they are generated in the stomach from other substances which contain the elements of starch or sugar. It is proper, however, that we should distinguish plethora from corpulence or adiposis. The former may consist in an excess of blood alone, or it may be associated with the latter. Adiposis, it is well known, may and does often exist when there is no real sanguineous plethora, in which case the food is improperly digested, leaving the fatty particles to be absorbed and deposited in the tissues (muscles, etc.), sometimes to the entire destruction of their integrity. In true sanguineous plethora, a low diet, or the prohibition of *meat*, soups, pastry, and certain vegetables, as beans, peas, etc., also such beverages as tea, coffee, brandy and other liquors, should be insisted upon. If the plethora be associated with adiposis, all carbonaceous articles of food, sugar, starch, etc., and malt liquors, should be strictly prohibited. If it is found difficult to subject the patient to these restrictions at her home, it will hasten the removal of the plethora if we place the patient in a water-cure establishment, or subject her to its processes under the care of an experienced nurse, whose duty it should be to regulate the diet as well as apply the proper baths, etc. The Turkish vapor bath has some reputation in England for the removal of plethoric conditions.

There are certain medicines which have a dynamic

influence over the circulation of blood to the extent of retarding the nutrition of the body. Dr. Rogers, of Michigan, asserts that the *veratrum viride,* in doses of five or ten drops of the first dilution several times daily, will tend to the arrest of the plethoric condition. If this should prove to be a fact, then we may presume that *aconite, gelseminum* and others of its analogues may have the same effect. It is supposed by some authorities that excessive water-drinking will have the result to diminish the amount of plethora, by acting as a diluent of the blood, but the value of this theory is more than doubtful.

In true corpulence, or adiposis, a somewhat different treatment is required. It consists principally in the withdrawal from the food of all starchy and saccharine substances, obliging the patient to live *almost wholly upon meat.* The advantages of this plan of treatment, with its successful results, is best set forth in a pamphlet by a Mr. William Banting, of England,[*] who, from being excessively corpulent, weighing 202 lbs, after living on the the following diet for one year, found his weight reduced 46 lbs, and his "girth" around the waist twelve and a half inches.

"For breakfast I take four or five ounces of beef, mutton, kidneys, broiled fish, bacon, or cold meat of any kind except pork, a large cup of tea (without milk or sugar), a little biscuit or one ounce of dry toast.

"For dinner five or six ounces of any fish except salmon, any meat except pork, any vegetable except potato, one ounce of dry toast, fruit out of a pudding, any kind of poultry or game, and two or three glasses of good sherry, claret, or madeira—champagne, port or beer, forbidden.

---

[*] "Letter on Corpulence." New York, 1864 (fourth edition).

"For tea two or three ounces of fruit, a rusk or two, and a cup of tea without milk or sugar.

"For supper three or four ounces of meat or fish, similar to dinner, with a glass or two of claret.

"For night-cap, if required, a tumbler of grog (gin, whiskey, or brandy without sugar), or a glass or two of claret or sherry."

It seems almost incredible that a man could actually get lean upon such a substantial, even luxurious, diet. Yet there is not wanting the testimony of other corpulent patients who have adopted this method, that it is quite effectual for the purpose, the mere abstraction of starch and sugar arresting the accumulation of adipose matter. Mr. Banting had previously tried "sea-air and bathing, much walking exercise, taken gallons of physic and liquor potassæ, riding on horseback, "lived upon sixpence a day, and earned it," if bodily labor be so construed, yet the evil still increased.

There are some medicinal measures which may be tried if the above diet does not succeed. *Liquor potassæ* has been useful in the removal of adiposis. This it is supposed to do by saponifying the fatty portions of the food before it has time to be absorbed. The dose is twenty to thirty drops, taken about two hours after meals. *Iodine, iodide of potassium*, and some other drugs, are alleged to have removed adiposis, but they are poisons, and should not be used. *Acetic acid*, when taken in the form of vinegar, is well known to cause a great decrease in size in fat persons, but in doing so it is likely to deteriorate the blood to a serious degree.

Quite lately a species of sea-weed known as the *fucus vesciculosus* has been highly recommended for the purpose of decreasing corpulence. It probably contains, in common with other sea-weeds, iodine in large proportion.

There is a condition of general *anasarca* which is often mistaken for plethora or adiposis, but may be distinguished by a careful examination. It may be allied to that condition known as leucocythæmia. This condition calls for such remedies as *potass. tart. et ferrum*, $\frac{1}{15}$ in grain doses, aided by *apis mel.*, *apocynum cann.*, *eupatorium pur.*, *china*, and *helonias*.

Anæmia, or Chlorosis.—These conditions should not be considered identical. The former generally proceeds from direct loss of blood, of or some of the vital fluids formed from the blood, to such an extent as to diminish the amount of the circulating fluid. The latter is generally a condition in which the quality of the blood is deteriorated, and is caused by deficient assimilation or depraved nutrition. The origin of the malady may exist in the digestive organs or nervous system.

It is therefore evident that the same treatment is not appropriate for both conditions. It has been, and still is to a certain extent, the opprobrium of the old school of medicine, that they prescribe ferruginous remedies indiscriminately in both diseases. The result is that only a portion of those who are thus treated are benefitted by the medicines used.

In uncomplicated *anæmia* iron is rarely indicated. It is only when the deficiency of blood is attended with a lack of certain vital constituents that this mineral should be used. For true anæmia the appropriate remedies are *china, helonias, aletris, hydrastis*. If chlorotic symptoms are associated with the anæmia, then it will be found beneficial to alternate with one of the above, *ferrum met.* or some of its preparations—I usually prefer the *pyrophosphate of iron*—or we may use such compound preparations as have been sanctioned by experience, whose constituents we find to be, by their pathogeneses, indicated in the case. Of this class of medicines, the *citrates of iron and quinia*, *elixir of bark and iron*, and some others, may be given with the best results. In true *chlorosis*, the most useful remedies are those which directly modify the digestive and nutritive functions, and increase the tone of the nervous system.

The medicines may be divided into three classes:

(a) *Ferrum met.*, or some one of its various preparations. It has been observed by nearly all practical physicians that *iron*, in its pure state, while it would restore some cases in a very short time, failed to improve the condition of others. But when used combined with some acid (as the *phosphoric*) or an alkali (as *potash*), it would act as a prompt remedial agent. The reason of this can best be understood by the Homœopathist, who finds that cases not amenable to *iron* alone, but which improved under *iron* and *phosphoric acid* (as *phosphate of iron*), presented symptoms which were not covered by the pathogenesis of either remedy singly, but by both completely. It is possible, even probable, that the alternation of the two medicines would effect the same curative result; but if, when given in combination, they cure promptly, we should not object to the form of administration.

(b) *Phosphoric, Nitric,* or *Muriatic Acids.*—Of these the *phosphoric acid* stands the highest in the estimation of the new school, probably because of its intimate relation to the nervous system. Here again, as with *ferrum met.*, we meet with cases wherein the acids alone will fail to effect a cure, but if we associate with the acid indicated, an alkali, or some other remedy which is also indicated, we speedily remove the disease. As examples of this rule, we often treat cases for which *phosphoric acid* and *calcarea* are the appropriate remedies. In such instances it does not matter whether we give the *phosphate* or the *hypophosphite of lime*, a curative result follows, because a diseased organism will appropriate to itself the proper curative agents, no matter in what form or chemical combination it is administered. I have found the *kali hypophos.*, *manganum hypophos.*, and *natrum hypophos.*, to be very useful in certain cases of chlorosis. *Natrum muriaticum* is an excellent remedy in this condition. Allopathic authorities speak of it as a powerful hæmatogen, in some cases nearly equal to *iron*.

(*c*) *Nux vomica, Ignatia, Strychnia.*—When chlorosis depends upon spinal exhaustion, or a want of tone in the nervous system, then this class of remedies are the only ones which will effect satisfactory improvement. Dr. Muller[*] reports many cases of chlorosis cured with *ignatia*, when all other means had failed. The *citrate of iron and strychnia*, or *strychnia* alone, has been found very effectual for the cure of chlorosis. The former preparation is a favorite one with me. In many cases of chorea, hysteria, and chlorosis, dependent upon spinal exhaustion, it has effected in my hands the most rapid and surprising cures.

Jahr[†] gives a long list of medicines indicated in chlorosis, of which the following are the most useful: *Calcarea, cocculus, ferrum, nitric acid, conium, pulsatilla, sulphur, china, platina, sepia and sulphur*, to which I will add *manganese, helonias, cimicifuga, senecio gracilis* and *aletris*.

In that analogous disorder known as leucocythemia, a somewhat different class of remedies are indicated,

---

[*] North American Journal of Homœopathy, vol. vi. p. 160.
[†] Diseases of Women.

namely, *ferrum iodatum, ferrum arseniosum,* and *helonias dioica* (also *plumbum*).

I hardly need add, that appropriate food, proper exercise, bathing, pure air, and healthful surroundings, together with the removal of all known causes of the complaint, whether mental or physical, are as important as the most carefully selected medicinal agents.

The Scrofulous Diathesis.—In those cases where we are convinced that the local disorders which threaten the life of the fœtus, or tend to arrest the progress of gestation, are due to scrofulosis, we must combat the diathesis with remedies adapted to each case, and at the same time prescribe such palliative remedies or topical applications as are indicated. A careful study of Hahnemann's chronic diseases is essential to the proper treatment of these cases. The remedies which are evidently appropriate, are those whose *local* effects upon the uterus, as well as their general symptoms, correspond most to the cases under treatment. The most prominent of these are the well-known anti-psorics—*Arsenicum, calcarea, conium, hepar sulph., iodine, lycopodium, mercurius, silicea, sulphur,* to which may be added other medicines equally efficient but less used—*Aurum mur., bromine, cistus canad., graphites, kreosote, lachesis, sepia;* or, *bromide of potash, bromide of iron, iodide of potash, iodide of iron, iodide of arsenic, iodide of sulphur, iodide of mercury, chloride of platinum, oleum jecoris aselli, chimaphilla, iris versicolor, phytolacca, rumex crispus, stillingia sylvatica.*

Return of Menstrual Crisis.—When, from previous habitual dysmenorrhœa, menorrhagia, or any undue tendency of blood to the uterus, or from any weakness or irritability of that organ, the menstrual nidus threatens to be so great as to threaten the continuance of pregnancy, such remedies should be selected as seem appropriate to each particular case, and should be administered during the *inter*-menstrual as well as the menstrual period. The medicines which will be most generally useful are—*Aletris farinosa, asclepias syriaca, belladonna, cimicifuga, caulophyllum, calcarea, gossipium, helonias, ignatia, platinum, pulsatilla, sanguinaria, secale, sabina, sepia, trillium, senecio gracilis, tanacetum;* or some one of the medicines mentioned as being capable of causing abortion.

But medicines alone are not capable of preventing the unnatural return of the menstrual nidus, unless we direct the woman to avoid undue exercise or warm baths, stimulant articles of food or beverages, coition, or any mental emotion of an unusual character, also any other influence which the physician with his knowledge of the idiosyncrases of his patient, shall consider inappropriate to her condition.

Zymotic Diseases.—In a work of this character and scope, the particular treatment of this class of maladies cannot be entered into. A mere mention is all that can be given to each disease.

(*a*) *Syphilis.*—For the specific treatment of this disorder the reader is referred to those works which treat of venereal affections, among which may be mentioned "Gollmann on Diseases of Sexual Organs," "Yeldham on Venereal Diseases," together with the papers on that subject to be found in our journals, etc. The best works of the dominant school may be consulted, the most scientific and rational of which is "Bumstead on Venereal."

There are certain remedies, however, which are not mentioned by any of the above authorities, but which have been found useful in the treatment of syphilis: they are,

Corydalis formosa, chloride of platina, iris versicolor, phytolacca dec., and *styllingia syl.* In my own practice these last-mentioned medicines have been prescribed with signal advantage, when the preparations of mercury, so much relied upon, were useless or nearly so. In addition to the above the *biniodide of mercury* and *iodide of potash* have been the most effectual preparations, which I have used in the treatment of syphilis.

(*b*) *Mercurialization.*—In the treatment of this form of drug poisoning, we must bring to bear upon the organism two forces: namely, the chemical and dynamic. These may be used singly or combined. It is well known to scientific medical men, that *iodide of potassium* actually enters into chemical combination with *mercury*, in the body—holds it in solution,—in which state it is carried out of the system through the various emunctories. For such purpose—it is needless to add—the drug must be administered in material quantities, or we shall fail to get any but its dynamic effects, which alone are not sufficient to eradicate the malady. *Chlorate of potash* and *hepar sulphuris* have a somewhat similar effect, but not to the same degree. If we wish to obtain a dynamic antidotal effect, or to remove certain local affections caused by *mercury*, we shall find most useful the remedies mentioned above, also, *aurum met.*, and *muriaticum, platinum chlor. phytolaca, podophyllum, nitric acid, iodine, mezereum, sulphur, styllingia, iris versicolor lachesis.*

In some instances it would be well to advise the patient to drink certain mineral waters, etc., abounding in *sulphur, iodine,* and other well-known antidotes to the effects of *mercury*.

(*c*) *Variola.*—The therapeutics of small-pox are so

well set forth in our standard works on Practice,* that no extended treatment will be given here.

My colleagues, with whom I have compared notes relative to the treatment of this disease, give, as the results of their experience, that *aconite, gelseminum,* or *cimicifuga,*† are the most valuable in the first stage of the malady; and in the second, *tartar emetic* or *variolin.*

My experience in small-pox has been limited. Two of my cases, however, were of the most severe character, (confluent) occurred in pregnant women, but they made a good recovery without any threatening of abortion. They were treated with *aconite, belladonna, variolin* and *tartar emetic.*

It is my impression that when abortions are reported to have occurred during an attack of variola, it is as often due to the harsh medication, as to the disease. At the same time it cannot be doubted that appropriate remedies may, by mitigating the violence of the attack, prevent the occurrence of miscarriage. Of these remedies none promise to be more efficient than *cimicifuga* (or *cimicifugin.*) Having a specific influence over the uterine-motor functions, it prevents the access of spasmodic or irritable conditions, which might otherwise obtain.

In relation to the extraordinary claims set forth for *thuja,* no positive testimony has ever appeared which in the least substantiated those claims. Those who are practically acquainted with the progress of a vario*loid,* know very well that at a certain period of the disease, the eruption, which appeared to be ripening, all at once is arrested, as it would seem from some powerful influence. The pustules suddenly "abort," a rapid recovery ensues and no pitting occurs. *This peculiar crisis never*

---

* Marcy and Hunt's Practice. † New Provings, page 106.

occurs in variola,—*i. e.*, in persons who have never been vaccinated. Now, in those cases in which *thuja*, 400th, is alleged to have been given with such splendid results, no mention is made whether they were variola or varioloid. This omission makes the testimony very unsatisfactory, if not entirely unreliable.

The same objections would hold good against the claims of the new remedy, *sarracenia*, were it not for the fact that this has been extensively tested in cases which were undoubtedly true variola. *Sarracenia* is strongly opposed by those who are sceptical of its powers. Since the publication of the first article treating of that plant,* in our literature, the testimony which has appeared, from public and private sources, relative to its prompt efficacy in arresting the course of the worst forms of variola, would place it at the head of all known remedies for that virulent malady.

Dr. Wilkinson, of England, in a recent work,† sets up extraordinary claims for the curative virtues of *veratrum viride* and *hydrastis* in this disease; but the same objections may be properly urged against the alleged value of those medicines. In the first, or febrile stage, the *veratrum v.* will undoubtedly alleviate the intense orgasm of the circulation, and in that way might prevent abortion. I do not think the claims set up for the *hydrastis* will ever be substantiated by positive experience.‡

(*d*) *Scarlatina.*—No physician who has witnessed cases of malignant, or even severe scarlatina, can doubt its influence for evil over the pregnant uterus and its contents. Although the cases are rare where scarlatina maligna occurs as late in life as the usual child-bearing

---

* "New Provings," page 384.
† "Small-pox—its preventive treatment," etc. London.
‡ Medical Investigator, January, 1865.

period, yet instances have occurred of abortion from that cause. When this result is to be feared, the most useful remedies are evidently *belladonna*, *apis mellifica*, *baptisia*, and *terebinthina*.

Several eminent physicians, among them Dr. Nankivel, of Penzance, and Dr. Blair, of this country, strongly recommend the *apis mel.* as a valuable remedy in malignant, as well as simple scarlatina. These recommendations are based upon successful clinical experience with this remedy. By reference to a former page it will be seen that *apis* has been alleged to cause abortion. This, together with its well-known action upon the kidneys, uterus, ovaries, etc., should commend it to us as an important remedial agent in cases of scarlatina where abortion threatens.

Belladonna, when the abortion threatens from arterial congestion, or spinal paralysis; *apis*, when the phlegmonous inflammation extends to the uterus, and the nervous centres are irritated; *baptisia*, when a profound typhoid and septic condition threatens; and *turpentine*, when the intestinal or urinary tract is the chief seat of the excessive irritation. Other analogous remedies—as *cimicifuga*, *sabina*, *caulophyllum* or *gelseminum* may be made use of, as auxiliary remedial agents.

In this disease, as in all others, of which mention will be made, while we should select the remedy in accordance with the law of *similia*, namely,—to correspond with all the symptoms and conditions of the patient,—we may select as an alternate remedy, another medicine which has a special affinity for the uterus, or, rather, has some decided influence in the production of abortion, while it is not Homœopathic to the *ensemble* of the disease under treatment.

Diphtheria.—The treatment of this malady[*] has been so fully set forth in the excellent lectures and monographs[†] of my western colleagues, that I will only add thereto such practical suggestions as are the results of my own experience.

The internal or constitutional remedies upon which I place most reliance are *baptisia, mercurius, bijodatus, phytolacca,* and *kali bichromatum.*

The topical remedies (which also act upon the general system), are *chlorate of potash, hydrastis canadensis,* and *per-manganate of potash.* I think the latter medicine was first used in diphtheria by myself. My first experience with it was in ulcers of an irritable character that had baffled other remedies both general and local. They healed kindly under a local application of the solution, and a consideration of the constituents of the agent led me to use it in diphtheria. I was struck with the prompt curative results obtained. No medicine with which I am acquainted so soon removes the exudation, *which does not return.* In this it has a great advantage over all others, for it is well known that the pseudo-membrane will return again and again under ordinary treatment with the applications in general use. Neither does the exudation extend—at least such has been my experience.

A glance at the chemical character of the remedy will explain its curative action.

"It may be made by mixing equal parts of very finely pulverized *deutoxide of manganese* and *chlorate of potassa,* with rather more than equal *parts of caustic potassa.*"[‡]

---

[*] Lectures on Diphtheria, by Professor Ludlam.
[†] Treatise on Diphtheria, by Professor Helmuth.
[‡] Parrish's **Practical Pharmacy,** p. 525.

We note that it contains *chlorine*, which is antagonistic to all zymotic poisons and septic conditions. *Chlorate of potash* and *muriatic acid* have been found the most useful remedies in diphtheria by physicians of the new and old schools of medicine. In the *kali per-manganatum* we have not only these, but also *manganese*, which, next to *ferrum*, is the most active and powerful hæmatogen known. Not only this, but it has a sustaining and tonic influence on the nervous system possessed by but few medicines. The *kali per-manganatum* acts not only as a disinfectant, and tonic, but as a *caustic*, and one, too, which causes little or no pain, no corrosion nor irritation. It may be applied in two ways, namely, (*a*) in strong solution with a camel's hair brush, and (*b*) in weak solution as a wash or gargle.

The maximum strength of the former preparation is ten grains to one ounce of distilled or pure rain water; of the latter, one drachm of the strong solution to one pint of pure rain water. No *alcohol* should be brought in contact with the drug, as a few drops will precipitate and render useless a large quantity of the solution.

The *fauces* should be thoroughly exposed, and all portions invaded by the exudation painted over with the strong solution; this should be done twice or three times daily, and the weak solution should be used as a gargle every three or four hours. None need be given internally in addition, as a sufficient quantity will get into the circulation to have its constitutional effect. It must be understood that I consider this medicine to act Homœopathically, but at the same time it may be said to act chemically, as will be shown in some future essay on this drug.*

---

* Two severe cases of diphtheria, treated mainly with the *kali per-manganatum*, will illustrate its efficiency:

(*a*) A young man who had been ill with fever for ten days; on the third

If I have somewhat stepped aside from my strict task, namely, the treatment of diphtheria in adult and pregnant women, my only apology is, that those remedies which have the greatest mastery over the disease at any age, will give us the greatest aid in the cases under consideration.

If premonitions of abortion appear during the course of a diphtheritic attack, we can appropriately alternate certain remedies, namely, *sabina, caulophyllum,* or *cimicifuga,* etc., with those which we are using for the original malady. The remedies mentioned will not retard the cure—in fact may assist our treatment, by their specific action upon the diphtheria itself, as well as upon the uterus.

Cholera.—The Homœopathic treatment of the different varieties of this disease, whether sporadic, as cholera morbus; or epidemic, as cholera Asiatica, will be found in the various text-books of our school, but especially in that admirable monograph* by a late lamented Homœopathician, Dr. B. F. Joslin, sen.

day I saw him; his fauces covered with a membrane, pearly at the edges, yellowish in the centre; breath very offensive; much debility; pains as if beaten all over; prognosis unfavorable. The weak solution was used every three hours, and *merc. bijod.*, 3rd, used alternately. In thirty-six hours all the membrane had disappeared; strength increasing; breath not offensive; discharged on third day of treatment.

(*b*) A child aged five years; two sisters had died of diphtheria within a week; tonsils coated with false membrane; breath offensive; very weak; croupy cough, etc. Used **the strong and weak tincture, alternate with** *phytolacca*, 1st. Cured in three days.

* Epidemic Cholera, 1852.

SECTION II.

(2) LOCAL OR ORGANIC.

(1) *Malformation of the Ovum.*
(2) *Malformation of the Membranes.*

As it is apparent that no remedial measures can prevent or remove any malformation of the ovum* or membranes, if we have a *suspicion* that such an abnormality exists, our only method of preventing abortion from such causes, is to prescribe such remedies as will tend to prevent the uterus from taking on an irritable condition, or, if such condition has already obtained, to remove it. In this way we may, if it is considered desirable, conduct the pregnancy to a termination at full time. It is doubtful, however, if such a termination is to be desired, for there are no dictates of humanity which can make it a moral duty for us to favor the birth of a monster, or deformed child, rather than its premature expulsion, unless the safety of the mother is to be gained by the former result. The medicines best adapted to bring about the uterine sedation mentioned above, are *belladonna, atropine, cimicifuga, caulophyllum, secale, sabina, tanacetum,* etc.

Placental abnormalities.—(*a*) Mal-location of placenta (placenta previa), (*b*) detachment of placenta. Obviously, no remedy can prevent the placenta from locat-

---

* Croserio (*vide* Obstetrics) intimates that the administration, to the mother, of such remedies as *sulphur, calcarea,* and other anti-psorics, in the 30th dilution, will remove diseased conditions of the ovum. (!)

ing over the os uteri, or any other portion of the surface of the uterus. But there are remedies which, after the mal-location has resulted in conditions and symptoms which tend to bring about an abortion, may so modify the condition, and hold the symptoms in abeyance, as to conduct the pregnancy to a favorable issue for both mother and child. If, when detachment occurs as the result of, or from falls, blows, or instrumental interference, hæmorrhage occurs, the patient is confined to her bed, kept cool and perfectly quiet, and such medicines administered as *arnica, hamamelis, hypericum, ruta,* etc., or any other remedy which seems specially Homœopathic to the symptoms, we may be gratified by seeing the hæmorrhage arrested, and with it the pain and other symptoms subside, and the patient ultimately recover. But such a favorable result is not usual. Generally, with the hæmorrhage, other symptoms, as chills, labor-like pains, etc., appear, which indicate that the blood is separating the membranes from the uterine walls. After this occurs, no remedial agent can arrest the abortion. Our sole effort should be to conduct it to a safe termination.

In cases of *placenta previa,* frequent hæmorrhages may occur, as the uterus expands and enlarges. These hæmorrhages will often subside spontaneously, at other times the loss of blood goes on until checked by appropriate remedies. In this form of hæmorrhage the blood escapes directly into the vagina, and does not permeate between the membranes and the uterus, hence the danger of an abortion is not as great.

The most efficient remedies to arrest the bleeding are *erigeron, erechthites,* arnica, hamamelis, aconite, millefolium, secale, sulph. acid, trillium,* etc., administered internally. Topically, we may use with advantage pieces of lint or cotton saturated with a solution of any of the above remedies, especially *erigeron, hamamelis*, and *arnica*. Cold water will sometimes arrest the bleeding, when applied to the os uteri. Care should be taken not to use too much lint, as it might act as a tampon, and cause the blood to collect and flow back, separating the membrane in the process, thus causing the very accident we are trying to avoid. We should also be careful not to irritate the vagina by unnecessary or harsh manual efforts, or we may set up sufficient reflex action to excite uterine contractions. The external application of cold water to the abdomen should be avoided, as it tends to the same result.

ORGANIC DISEASE OF PLACENTA.

(*a*) *Fatty Degeneration.* (*b*) *Calcareous Degeneration.*
(*c*) *Hydatid Degeneration.* (*d*) *Molar Degeneration.*

In the present state of our therapeutical knowledge, we know of no remedy or remedies which could prevent the occurrence of the above organic changes. These diseases of the placenta probably have their origin in some dyscrasia pervading the blood of the mother. This dyscrasia is generally the psoric or the syphilitic.

The science of diagnostics has not reached that point where we can decide if either of the above changes have taken place in the placenta. It is true we may safely predict the existence of a hydatid mass in the uterus, if individual hydatid vesicles have been discharged per vaginum; but we cannot safely give an opinion as to the existence of a mole, until the uterus has expelled it. If a woman has had conceptions which have resulted in such organic diseases, as was proved by the expulsion of the degenerated mass, we may *fear* the occurrence of similar changes in the placenta, as a result of subsequent conceptions, especially if the symptoms of the pregnancy are peculiar and unnatural.

The *treatment* of organic diseases of the placenta; or rather the *preventive* treatment of abortion from such causes, is manifestly obscure.

Dr. Tyler Smith says—"When the fœtus is threatened with death because the placenta cannot perform its nutritive and respiratory functions, we may, through the mother, act upon the placenta, and assist in the performance of its functions."* Dr. Power prescribed the inhalation of air containing an increased quantity of *oxygen*, or the use of medicines containing a large proportion of *oxygen* in a loose state of combination, as *nitric acid*, in cases where the child has been lost repeatedly in the latter months of pregnancy. Dr. Simpson states that he has found *chlorate of potash* useful in cases where the fœtal respiration is imperfect. "In the prevention of abortion from *fatty degeneration* of the placenta," Dr. Smith says,† "the strength of the mother should be supported in every way. The *chlorate of potash*, *nitric acid*, mild preparations of *iron*, and, above all, fresh *air*, should be recommended. The treatment of fatty placenta should be the same as fatty heart, or fatty degeneration of any other organ."

The *Homœopathic* treatment of fatty or calcareous degeneration is yet uncertain. We know of no medicines which cause these organic changes.

If we suppose, from the patient's previous history, and the present symptoms, that the placenta is undergoing degeneration, we may administer medicines for the purpose of acting through the maternal circulation upon the diseased organ.

* Lectures on Obstetrics, page 201.   † *Ibid.*

The anti-psorics may be consulted, for in that class we shall be most likely to find the appropriate remedy.

For *fatty degeneration*, such remedies as *baryta carbonica, thuja, kali bromatum* or *nitric acid,* may be used.

For *calcareous degeneration*, we can suggest no specific remedy; nor for the hydatid or molar; the general symptoms of the patient must be our guide in the selection of the medicines.

If we become satisfied that these organic changes have taken place, it seems evident to me that our duty should be to rid the woman of them as soon as possible. To this end, we may separate the degenerated mass from the uterine walls by the careful use of the sound or flexible catheter, or by the injection of tepid water; and in case of hydatids we may break down the accumulation and separate the mass from the uterus by the same means. Dr. Bedford relates the case of a woman in extreme danger from loss of blood, in which he successfully broke down the hydatids with a female catheter. Dr. Tyler Smith says the hydatids are sometimes expelled with difficulty, and he once saw a case in which the uterus had been ruptured by the violence of its contractions in expelling an hydatid mole.

Several cases of hydatids successfully treated have appeared in our periodicals; one by Dr. Comstock,[*] and another by the writer.[†]

The after treatment is the same as in cases of actual fœtal abortion.

---

[*] Medical Investigator, vol. ii., p. 10.
[†] American Magazine of Homœopathy, vol. 1.

SECTION III.

Reflex (Exciting).

1. CENTRIC.

(*a*) *Emotions of Fright, Anger, Grief*, etc.

We find in our repertories the following remedies recommended for the consequences of the above mentioned emotions. I have placed those in ITALICS which are most likely to be useful in abortion excited by such causes.

ANGER.—*Aconite*, bryonia, *chamomilla*, colocynth, *nux vomica*, *platina*, staphisagria.

FRIGHT.—*Aconite, belladonna, gelseminum, hyosciamus, ignatia, lachesis*, opium, *pulsatilla*, sambucus, *veratrum*.

GRIEF.—*Ignatia, phosphoric acid, hellebore.*

JOY.—*Coffee, scutellaria, cannabis indica.*

Most of the above medicines are recommended from general indications, and no true Homœopathician would prescribe any one of them unless the general symptoms corresponded with those present in the attack. If any other medicine in our extensive Materia Medica seems particularly indicated, no attention should be paid to the emotional cause, which should not be deemed important except in the absence of characteristic symptoms.

A case in which this latter condition obtained, once

came under my care. In the absence of any guiding symptom of importance, *phosphoric acid* and *ignatia* were prescribed and prevented the abortion. Disappointed love was the cause of the threatening symptom.

(b) *Direct blows upon the brain or spinal cord.*

The remedies most appropriate to meet the effects of concussion of nerve substance, are *arnica, conium, cimicifuga, cicuta, gelseminum, hypericum, mercury, nux vomica, opium, quinine, rhus tox,* and *sulphuric acid.*

It is needless to give the special indications for each of the above remedies. The specific curative power of *arnica* is well known to all practical physicians; it should be used freely externally upon the portion which has received the injury, and at the same time administered internally; *conium* is indicated more particularly for the chronic effects of concussion. *Cimicifuga* will often be found very useful, especially after concussions, when, with the uterine pain, severe cephalagia is present (also *quinine*). *Gelseminum* is often useful when a condition of general paralysis of the voluntary muscles, with loss of sense, and blindness occurs.

Hypericum is said by some physicians, who have tested its virtues, to be more useful in injuries to nervous tissues than any other medicine.

Ignatia and *nux vomica* when tetanic symptoms are feared.

Opium, when we find sopor, or coma, and symptoms similar to apoplexy.

Sulphuric acid, when the [only] notable symptom is sudden and excessive uterine hæmorrhage.

(c) *Medicinal, namely*—Quinine, Nux vomica, Gelseminum, Cimicifuga, Secale, Apis (?) *and others.*

If we know, or have reason to suspect, that the impending abortion we are called upon to avert has been excited by any of the above medicines which act through the nervous centres, we must, if called in season, attempt to empty the stomach of the poison. If it has already entered the circulation, we must use those general and special antidotes which are at our command.

The best known antidotes against the immediate effects of *quinine* are—*Morphine, arnica, cimicifuga, ipecac, pulsatilla,* and *veratrum viride.*

Nux vomica (Strychnia).—Administer *gelseminum* in appreciable doses, and place the patient under the anæsthetic influence of *chloroform* until the spasmodic (tetanic) symptoms cease.

Gelseminum.—Give stimulants, brandy, whiskey, and *ammonia,* with *nux vomica,* or *arnica.*

Cimicifuga.—The antidotes of the *black cohosh* are *atropine, opium,* and *secale,* in minute quantities.

Secale.—For the effects of *ergot* on the uterus, I have found *caulophyllin* and *atropine* excellent antidotes.

Apis mel.—*Ammonia,* in stimulant doses, is the best medicine to ward of the bad effects of the bee-poison upon the nervous centres.

II. CONCENTRIC.

(a) *Parotidean Irritation.*

By referring to a former page of this work, it will be seen that in one case at least parotitis was a cause of abortion. To prevent such a result from metastasis of mumps, we should prescribe about the same remedies as in case of threatened ovaritis or orchitis from the same disease, namely, *apis mel.*, belladonna, *pulsatilla*, or mercury. The two in *Italics* are probably the most useful, and would be indicated in the metastasis referred to.

(b) *Thyroidal.*—If the reflex irritation which has located in this gland has been diverted from it to the uterus, directly or through the mammæ or ovaries, the remedies to be mentioned hereafter will be indicated. The treatment of goitre with external applications of *iodine, merc. bin.-iod.*, or *bromine*, should be suspended during pregnancy, and small doses of these remedies used instead. These minute doses, while they do not divert the irritation, act as a sedative or palliative to it, and really prevent such irritation from impinging upon the uterus. If uterine pain, etc., has already set in, we must select the remedy in accordance with the general principles of our school of medicine. This latter observation will apply to all the following causes, and need not be hereafter repeated.

(c) *Mammary.*—If abortion or premature labor threatens from prolonged lactation, the child should be immediately taken from the breast, and lotions of *belladonna, aconite,* or *camphor*, applied to the glands, to arrest the secretion of milk and subdue the existing irritation. If the cause be mammary abscess, the same lotions should be applied, also *phytolacca*, which is an admirable remedy in this affection. At the same time *belladonna, apis, phosphorus,* or *phytolacca*, should be administered internally. If these agents do not bring about speedy resolution, and suppuration is inevitable, place the patient (if she is nervous and sensitive) under the influence of *chloroform*, and open the abscess, after which apply soothing applications, such as *calendula*, or *poppy leaves*, or lotions of *aconite, calendula,* or *hamamelis*. The pathogenesis of *cimicifuga* seems to intimate that it may be useful in abortion from mammary irritation.

Scanzoni causes premature labor by applying dry cups to the mammæ. Dr. Bedford admits that abortion may be caused by irritating these glands; yet we are told by some late medical writers that habitual abortion or sterility may be prevented by applying a child to the breast during the period when the usual menstrual crisis returns!

We are even advised by certain obstetric authorities to apply the child early to the breast, in order to avert a metritis, uterine congestion after confinement, or puerperal fever.

It would seem by the above that mammary irritation acts both as a cause and as a preventive of abortion and uterine irritation.

Would it not be proper and useful then, in cases of threatened miscarriage, to apply *mildly* irritating substances to the mammæ? Weak sinapisms of *mustard, capsicum,* or *arnica,* a dry cup to each breast, or even a warm emollient poultice, might divert the irritation from the uterus to these organs, and thus avert the loss of the embryo, or a premature labor.

(*d*) *Dental.*—If we have to treat a merely functional disorder of the trifacial nerves, or any nerve branches which supply the teeth, the remedies will be mainly, *aconite, belladonna, atropine, china, quinia, valerianate of zinc, coccionella, gelseminum, arsenicum, spigelia,* and *pulsatilla.*

If the affection is organic, namely, carious teeth, the remedies most likely to palliate the pain and reflex irritation, when given internally are *mercurius, nitric acid, antimony, kali hyd., phosphorus, manganese*, or those above mentioned.

The repertories decree that for "odontalgia" in pregnant women, we should give *belladonna, calcarea, manganese, nux moschata, nux vomica, pulsatilla, sepia,* etc.

We are cautioned against the extraction of carious teeth during pregnancy. But in many cases the irritation caused is so intense, and so intractable to medicinal treatment, that it is better to risk the operation. By the aid of *chloroform*, however, we can generally extract the tooth and avoid the risk of abortion from the irritation attendant on the operation. If it should be objected to, or feared, we may resort to topical, palliative measures, such as cotton wet with *aconite, gelseminum*, or *opium*, and applied to the exposed nerve; or we may use *kreosote* in the same manner. I have often used with permanent benefit a plugging composed of *aconite, chloroform* and *copal*, which formed a solid, impervious, and lasting compound.*

The usual operation of filling, plugging, etc, of the dentist, need not be deferred, unless we have special reasons for so doing.

Odontalgia is often accompanied by inflammation of the buccal cavity, stomatitis, gingivitis, and other diseases of the mouth, nearly all of which will subside under the specific curative action of *chlorate of potash*, used in the form of a lotion.†

Ulcerative affections of the teeth demand the same remedies advised for pain in carious teeth, and in addition, *hepar sulph.* and *phytolacca*. It is better to extract the tooth than to allow the abscess, if deep, to go

---

* ℞ Aconite tincture, ʒj   † ℞ Chlorate of potash, grs. x.
  Chloroform, ʒj   Water, ℥4
  Gum copal, grs. x.

on to suppuration with the intense suffering which sometimes accompanies it.

(*e*) *Gastric.*—The variety of gastric irritation which usually causes abortion, is obstinate and excessive vomiting. This may begin at or before the sixth week, and continue until the period of confinement. It is not only often obstinate, but intense and painful, and even accompanied with severe gastric irritation, simulating acute gastritis, **or ulceration.** I have seen patients, victims to this disorder, who became debilitated and emaciated to the last degree from the inability to retain the least food or beverage upon the stomach. The vomiting, in such cases, was accompanied by the most distressing pyrosis, burning pain in the stomach, and spasmodic, empty retching, **or with the ejection of bloody, grumous matters, and even pure blood.** But it is a strange fact, that even in these cases of extreme severity, pregnancy will sometimes go on **uninfluenced, and confinement** occur at its proper period. In other instances, however, less severe, and even quite mild, abortion will occur, apparently from no other cause than the efforts at vomiting.

Tyler Smith says—"Women who have undertaken sea-voyages during pregnancy, have died from **the combined effects of sea-sickness and the vomiting of pregnancy.** * * In the worst cases, **women who are not relieved, or who do not abort, perish slowly from starvation,** or they die from the rupture of a blood-vessel, convulsions, or exhaustion, after violent and continued fits of vomiting." It is in these severe cases that it is proper and humane to arrest the suffering of the **patient** by inducing an abortion by artificial means; **but of this** we shall speak further on.

This affection, which, **at first** sympathetic, or reflex,

often becomes a local disease, is quite amenable to Homœopathic *treatment*. In the simplest form, namely: morning vomiting—a dose of *nux vomica* or *podophyllum* in the evening, with *ipecac* or *cimicifuga* in the morning before rising, will often palliate, if not correct the symptoms. When the vomiting occurs after meals, and the food is ejected,—*pulsatilla, nux vomica, iris v.* or *ferrum*. If the vomiting occurs at all periods, and is aroused by the least food and drink—*arsenicum, euphorbia cor., veratrum alb.,* or *viride, kreosote, sanguinaria* and *oxalate of cerium* are useful. Of these, *arsenicum* and *sanguinaria* are indicated, if there is great thirst and burning pain in the stomach, vomiting of bloody mucus, etc.; *veratrum* and *arsenicum*, if the prostration is excessive: *kreosote* has been found almost specific, by the old school, in many severe cases—in minute doses, it has been found useful by physicians of the Homœopathic school. The *oxalate of cerium*, first recommended by Professor Simpson, I have prescribed with great benefit in the first decimal trituration in some very severe and obstinate cases.

Other remedies have occasionally been used with success when ordinary remedies fail. Among them are—*tincture of aconite,* in drop doses; *colombo* and *wild cherry,* in infusion; *chloroform,* five or ten drops in a spoonful of *mucilage; nitrate of bismuth,* when the acidity and epigastric pressure is excessive; *salicine, helonias* and *nitric acid,* "if the stomach be in an atonic condition."

(*f*) *Rectal.*—The rectal causes of abortion—namely, dysentery, hæmorrhoids, etc., have been described.

The remedies which have been found most useful in *dysentery* are—*aloes,* aconite, *baptisia,* colocynth, *ipecac,* iris versicolor, *mercurius,* nux vomica, *podophyllum.*

Those in *Italics* would be most applicable in cases when abortion was threatening, in consequence of the sympathetic or reflex irritation. If no one of the above remedies covered the *ensemble* of the symptoms, some other remedy not capable of causing dysentery may be alternated with the one selected out of the above list. Thus, *aconite* and *caulophyllum*, or *colocynth* and *aletris* could be alternated. The symptoms of *sabina*, however, cover nearly the whole ground, for that medicine not only causes abortion, with inflammation of the urinary organs, but it has enteritis and dysentery, bloody stools, etc., among its pathogenetic effects. *Asarum europeum*, which causes abortion, has been found very useful in acute dysentery. The same is true of *turpentine*. These last remedies should not be forgotten when we are called to treat this disease in pregnant women.

The topical treatment of dysentery during pregnancy should not be omitted. By applying some medicine locally to the rectum, and causing a sedation of its nerves, we shall also calm the irritated nerves of the uterus. The remedies best calculated to subserve this end are—*aconite, gelseminum, hamamelis, opium, morphine, atropin*. These medicines may be added to mucilaginous preparations, starch-water or pure water, in proper proportions,* and thrown into the rectum, where it should remain. I am confident that I have in one instance prevented abortion by the use of an enema of ten drops of *gelseminum* to one ounce of starch-water; and in another prevented a premature labor by the use of half a grain of crude *morphine* in two ounces of pure water. In each case the severe labor-like pains ceased in

---

* Aconite, Gelseminum, ⊙, 10 drops to one ounce of water.
Opium Tinc., ⊙ (Laudanum) 20 drops to one ounce of water.
Hamamelis, ⊙, 1 drachm to one ounce of water.
Morphine, 1-10th, Atropin, 2d, two or three grains to one ounce.

half an hour after the enema was given. In some cases a suppository of one grain of pure *opium*, introduced into the rectum, will have a prompt effect in calming the tenesmus and labor-like pains.

In case the uterine irritation proceed from *hemorrhoidal* inflammation, the following remedies will be found applicable:

Aconite, aloes, belladonna, æsculus hip., collinsonia, hamamelis, iris versicolor, mercurius, nitric acid, nux vomica, podophyllum, sulphur, turpentine.

In this disease, as in dysentery, local applications should not be neglected. Lotions or cerates* of *atropine, hamamelis*, or *opium*, may be applied to the inflamed and protruded tumors, and the patient advised to assume and maintain the recumbent position, until the local and reflex irritation has subsided.

Fissure of the anus, one of the most painful and intractable local affections with which we have to deal, has been known to cause abortion. The reflex disturbance which this apparently insignificant lesion can produce, is quite astonishing. The remedies Homœopathic to this, are *ignatia, nitric acid, plumbum, arsenicum, sulphur*, etc. The most notable cures have been made with *nitric acid*, 30th, and *ignatia*, 30th. I have, however, cured two severe cases—one in a pregnant woman, with *nitric acid* and *ignatia* (at the 3rd dilution). An enema of *nitric acid* was used at the same time with the internal treatment. (Ten drops of *nitric acid* [dilute acid] was added to four ounces of water.)

Diarrhœa.—That variety capable of causing miscarriage, will be best met by such remedies as *arsenicum*,

---

* *Cerate* of Atropine, 1 grain to 1 ounce,
" Opium, 10 grains to 1 ounce,
" Hamamelis, 1 drachm to 1 ounce.

euphorbia cor., *ipecac*, *iris versicolor*, *mercurius*, *podophyllum*, *cuprum*, *pulsatilla*, *sulphuric acid*, *veratrum album*, and *veratrum viride*.

Constipation, or "functional impediment of the bowels," should be relieved by the most prompt and efficient measures. In order to effect this, purgatives should never be resorted to, for they will only intensify the difficulty, and aggregate the tendency to uterine irritation. The diet of the patient should be made as laxative as possible. The free use of fruits, corn bread, cracked wheat, berries (particularly whortleberries) should be urged, and one of the following remedies prescribed, namely, *arsenicum*, *bryonia*, *collinsonia*, *mercurius*, *nux vomica*, *plumbum*, *platina*, *lycopodium*, and *sulphur*.

The persevering use of enemas are particularly recommended. They may be composed of pure water, or water holding in solution common salt, ox-gall, soap, or molasses—in quantities not sufficient to be stimulating. Decided benefit has resulted from enemas medicated with small quantities of *nux vomica*, or some other medicine Homœopathic to the local condition of the intestines. If violent purgatives are the cause of the symptoms of abortion we are called to treat, they are to be met by the same means advised for the relief of a dysentery or diarrhœa.

(*g*) *Vesical.*—The causes of abortion which emanate from the urinary organs are not unimportant, and any unusual irritation of the bladder or urethra, during pregnancy, should be met with prompt remedial treatment. The kidneys are not so powerful to originate reflex irritation as the other portions of the urinary tract.

Acute cystitis calls for *cannabis*, *cantharis*, *apis*, *aconite*, *senecio aureus*, *mercurius*, *pulsatilla*, and *turpen-*

tine (European physicians estimate highly the curative powers of *pareira brava*, and *buchu*, in this affection).

Chronic cystitis demands *copaiva, chimaphilla, erigeron, erechthites cubeba, thuja, turpentine, eupatorium purpureum*, and *hydrastis*.

Nephritis, and *urethritis*, acute and chronic, demand the same remedies. For special indications, refer to "Materia Medica," "New Remedies," etc. The free use of diluent and mucilaginous drinks should not be omitted, as they greatly aid the action of specific remedies. Among the most useful of these semi-medicinal agents, are *ulmus fulva* (slippery elm), *galium aparine* (cleavers), *Althæa off.* (marsh mallows). These should be prepared in simple infusion with cold or warm water, and drank *ad libitum*, or in given quantities (as a wine-glassful), at given intervals, alternately with the medicinal remedies.

(*h*) *Vaginal.*—The causes of abortion originating in the vagina, are acute and chronic inflammation, malignant disease, and granular vaginitis.

Acute vaginitis, occurring during pregnancy, is an important disease. West has known it to extend to the uterus, producing metritis. The swelling, heat, pain, and irritation, have been known to form an exciting cause of miscarriage. When attended with *pruritus* of the vulva, it is a most distressing affection, producing intense irritability of the nervous system. This latter symptom generally attends an aphthous condition of the vulva and vagina. The disagreeable itching is felt not only in the vulva, but in the internal organs.

The *treatment* of vaginitis (simple acute) consists in the internal administration of *cantharis, cannabis, mercurius, pulsatilla, thuja, sabina*, and such auxiliary measures as cool hip baths, emollient and soothing enemata. The injections which I have found most useful are infusions of flaxseed, poppy-leaves, calendula flowers, and slippery elm. There are other forms of vaginitis which require different medicines. In the erysipelatous, or erythematous variety, when the inflammation extends to the cellular tissue, with extensive swelling, and even suppuration, it will be necessary to give *belladonna, apis, coccionella, mercurius, collinsonia, cantharis*, and *aconite*.

Granular vaginitis requires *mercurius, phytolacca, sabina*, and *cantharis*.

Vesicular vaginitis—*rhus tox, commocladia, petroleum, cantharis, croton tig.*, and *dulcamara;* and the *pustular* variety—*tartar emetic*, and *sabina*. The *gonorrhœal* has been mentioned **under that head, and the** *chronic* under " Leucorrhœa."

If *pruritus* attends either of the above varieties, some one of the above medicines will be indicated, especially *cantharis, sabina, collinsonia*, or *platina*, but topical applications cannot safely be dispensed with. For the relief of this distressing symptom we have not found anything so useful as the "lotion of *borax* and *morphia*,"* applied to the vulva, or used as a vaginal injection—one ounce every six hours. Dr. West† says, "nothing relieves the *pruritus* which accompanies the decline of vaginitis more than *goulard water* **and** *hydrocyanic acid*, in the proportion of two drachms of the latter to eight ounces of the former" These applications might, with some reason, be objected to, if the disease occurred in the unimpregnated state; but when we wish to alleviate

---

* Sodæ Sub-boracis ℨ ix,
 Morphia Sulph., gr. viii.,
 Aqua Rosæ ℥ x.

† Diseases of Females, p. 480.

promptly a symptom likely to result in serious consequences, we must not adhere too strictly to purely Homœopathic remedies, given internally. Other applications may be found useful, such as *aconite-water, eupat. arom., hamamelis, chloroform-vapor.*

Vaginismus is most successfully treated by the administration of *gelseminum, belladonna, atropin, hyosciamus, ignatia,* or *platinum.* This symptom is generally found in hysterical women, and the remedies used for that condition will be useful. Injections of water, medicated with one of the above remedies, will aid the cure, or a soft cerate, in which they are incorporated in proper proportion,* is sometimes still better. If the vaginal irritation has been mechanical, remove the source of irritation, and use mucilaginous injections, or enemas of *aconite, hamamelis, opium* or *arnica,* in the proportion of one drachm to one pint of water.

Hysterical.—This condition of the system may be kept in abeyance, and the attacks warded off or palliated by the use of *aconite, gelseminum, atropine, ignatia, platinum, cimicifuga, caulophyllum, nux moschata, ambergris, zincum val., scutellaria,* or *cypripedium,* also the use of remedies calculated to remove the pathological state upon which the hysterical irritation depends.

Epilepsy, though generally incurable, may be palliated by *atropine, hyosciamus, belladonna, stramonium, zincum phos.,* and many other remedies mentioned in our Materia Medica and late Journals.†

Falls, Jumping, Blows, etc.—These accidents, when

---

* Gelseminum (tinc.)   ʒi to ℥i
  Belladonna (ext.)   ʒi to ℥i
  Atropin,   gr.i to ℥i.

† North American Journal of Homœopathy, vol. 12, p. 261.

acting as concentric causes of abortion, require about the same remedies as when acting centrically.

If the patient receive a blow, or has a fall, and the injury is followed by any pain or soreness in the uterine region, or flow of blood from the vagina, she should be required to assume the recumbent position, and enjoined to keep very quiet. At the same time we should prescribe *arnica, rhus tox, hamamelis, sulphuric acid, cinnamon, erigeron*, or *caulophyllum*, as the nature of the case and symptoms demand. Allopathic physicians place much reliance, in cases of impending abortion from injuries, upon the free use of *morphine*. They prescribe a " full dose," from one-eighth to half a grain, or more, if the nervous excitement is great. This treatment is not more at variance with strict Homœopathic practice, than is the administration of *morphine* after fractures or severe surgical operations. It deadens sensation, relieves pain and excitement, paralyzes the motor actions of the womb, and places the system in such a condition (in *splints*, if the phrase is allowed), that the injury is powerless to excite the amount of reflex irritation necessary to cause abortion. I have never resorted to *morphine* but in one instance, and the patient escaped a miscarriage; in all others, I have relied upon other remedies, with varying success.

SECTION IV.

FUNCTIONAL DISEASES OF THE UTERUS.

Congestion of the Uterus.—The treatment of this condition will depend on the nature of the congestion, whether acute or chronic. The remedies recommended in the Repertories for " congestion of blood to the uterus" are

Belladonna, bryonia, china, crocus, hepar sulph., mercurius, nux vomica, platina, sabina, secale, sulphur.

Of these, *belladonna, crocus, sabina* and *secale,* are the most likely to be indicated in abortion from this cause.

But there are other remedies equally important in the treatment of congestion and its consequences. It should be borne in mind that *any remedy capable of causing abortion, will cause congestion of the uterus.* We have, therefore, a large number of remedies from which to choose.

For *acute* congestion the following are the most efficient—*veratrum viride, sabina, belladonna, cimicifuga, secale, caulophyllum, tanacetum, aletris, mitchella* and *terebinth.*

For *chronic* congestion—*china, sepia, platina, secale, sulphur, trillium, nux vomica, helonias, murex, ferrum* and *calcarea.*

In peculiar cases, *sabina, cimicifuga, aletris, tanacetum* and *terebinth,* may be Homœopathic in chronic (passive) engorgements of the womb, by virtue of their secondary action.

Engorgement, Induration or Hypertrophy of the neck of the womb, like congestion of the body of that organ, may be acute or chronic. In cases of *simple engorgement,* without ulceration, abrasion, or œdema, we shall find the following remedies, aided by enemas of cool aconite-water, or calendula-water quite sufficient for its removal—*belladonna, cimicifuga, mitchella, platina, sabina* and *aquaphobin.*

In *induration* or chronic inflammation, we shall find most useful—*sepia, murex, platina, mercurius iod., kali brom., kali hydriod, conium, china, sulphur, helonias, stillingia,* and many others.

For œdematous engorgement, or hypertrophy, the the most appropriate remedies are—*arsenicum, conium, apis mel. arnica sepia, iodine, china, helonias,* etc.

The internal treatment of the last two conditions of the cervix, may be materially assisted by the topical use of the same medicines, but especially *arnica, iodine, conium, bromide of potash* and *china.* These may be used, largely diluted, as enemas, or in some cases the stronger preparations may be applied to the cervix, by means of a soft brush.

Leucorrhœa.—The treatment of this affection depends so much on the *character* of the discharge, the pathological condition to which it owes its origin, and the symptoms with which it is attended, that it is absolutely necessary the physician should consult his Materia Medica, or some elaborate work like Jahr on Diseases of Women. It would not be proper, in a work like this, to enter fully into the special indications of each remedy. The best that can be done is to designate those remedies which will be found most useful in the varieties of the disorder.

In *mucous* or cervical leucorrhœa, or that variety

which is secreted by the mucous follicles of the cervix uteri, the most appropriate remedies are—

In *acute* cases—*pulsatilla, senecio gracilis, sabina, mercurius, bovista, cantharis, podophyllum, dulcamara, senega, kali hyd.*

In *chronic* cases — *pulsatilla, stannum, sulphur, conium, sepia, copaiva, merc. iod., kali bromatum, phytolacca, trillium, thuja, calcarea carb.*

In cases when the discharge has produced abortion or erosion of the os uteri—*nitric acid, merc. cor., merc. iod., kali hyd., kali brom, arsenicum, lachesis, phosphoris, sabina.* (See Ulceration of Os Uteri.)

Where there is anæmia or debility—*china, helonias, ferrum, manganese, hydrastis, aletris, phosphoric acid,* and the *hypo-phosphites.*

In cases of chronic cervical leucorrhœa, it is useful and proper to use topical applications. Any remedy Homœopathic to the disorder may be applied locally to the canal of the cervix, by means of a syringe of peculiar shape, and with the result of hastening the cure. Those caustic substances which act Homœopathically upon diseased surfaces may be applied in the above manner, or by means of a brush, or even in substance. Of these the most useful are *argentum nit., kali permanganatum,* and *nitric acid.* The topical treatment here advised is not to be used during, but *previous* to pregnancy.

The treatment of a leucorrhœa should be conducted upon the same general principles as an acute coryza, or catarrh of the air passages, or profuse secretion from glandular or mucous surfaces. It will be noticed by the careful student of Materia Medica, that any remedy capable of causing a peculiar mucous discharge from the nose, is capable of causing a similar, or even identical discharge from the bronchia, bladder, intestines, and cervical canal of the uterus. If the discharge from the nose be acrid and corrosive, the discharge from the cervix will have the same irritating quality: if tenacious in the larynx, it will be tenacious elsewhere; if green, yellow, or white in one place, it will have the same color elsewhere. The idea of any remedy causing specifically a mucous discharge from one surface, and not from any other, is absurd and unphilosophical. If the physician will examine any complete pathogenesis, he will be convinced of the truth of this assertion. Medicines have a specific affinity for *tissues*, not *organs*. This is a great truth, and one of immense importance in practice, and a valuable guide to our study of the science of Therapeutics.

Vaginal or Epithelial Leucorrhœa.—Owing to the total absence of a microscopical examination of the leucorrhœal discharges, occurring during a proving of our medicines, we cannot designate the remedies indicated for a strictly *epithelial* discharge; nor can we designate with any more certainty the proper remedies for a *purulent* leucorrhœa, although Jahr indicates *cocculus, mercurius, sabina, china, sepia, copaiva,* and *nitric acid.* I do not suppose *cocculus* or *china* capable of causing a purulent discharge from any tissue. It is not in the nature of the medicine; and I suspect if the original provings were examined, they would be found faulty in that respect. It is probable that *china*, by its Homœopathicity to extreme debility, will cure purulent discharges, for it is now known that anæmia will cause a mucous discharge to become purulent, and a removal of the debility will cause the pus to disappear. *Cocculus* acts principally upon the nervous centers, and it is difficult to imagine how it can cure a purulent discharge.

For *epithelial* leucorrhœa I would *theoretically* designate—*dulcamara, borax, bovista, calcarea, pulsatilla, sulphur, kali hyd., merc. iod., senecio, cannabis, copaiva, cubebs, euphrasia* and *sticta*.

For the different varieties of *vaginal* leucorrhœa, I would refer the reader to Jahr,* who gives minutely the symptomology of each remedy likely to be useful in leucorrhœa generally.

When it arises from simple, acute catarrhal vaginitis, *dulcamara, pulsatilla, kali hyd., sticta, senecio gracilis* or *merc. iod.* will generally remove it.

If from *aphthous* inflammation, *borax, caulophyllum, sulphur* and *muriatic acid* will be found most useful. On a previous page this form of vaginitis and its peculiar discharge has been mentioned, also the distressing pruritus to which it sometimes gives rise, and that it is apt to occur during pregnancy, and be a cause of abortion. The two remedies having greatest control on this variety of the disease happen to be very different in constitution. One is a vegetable, the other a mineral substance, yet both exercise a similar effect upon the organs of generation, as mentioned in the "Medicinal Causes of Abortion." Both cause labor-like pains, and are capable, under certain circumstances, of causing abortion and premature labor. Both are Homœopathic to, and curative in, aphthæ of any mucous surface. It follows that no remedies are likely to be so useful in threatened abortion from aphthæ and pruritis, as *borax* and *caulophyllum*.

Sulphur, and even *sulphuric acid*, according to Hartman,† are both useful in aphthæ, but more against the constitutional diathesis, than the immediate attack and its consequences.

---

* Diseases of Women.   † Diseases of the Children.

Muriatic acid, says Teste, is the specific remedy for that form of aphthæ, termed by the French writers *muguet*, and which may sometimes attack the vagina, especially in very low, atonic states of the system.

The local application of *borax* has been mentioned under "Pruritus." *Caulophyllum* may be used as an injection—the tincture in water. *Sulphurous acid* is promptly curative in the form of enema, when there exists in the vaginal discharge the fungi peculiar to some aphthous diseases.*

If the whole vaginal surface has become involved, and by means of the speculum we discern that the epithelium has nearly or wholly disappeared, leaving the villi exposed, the villous coat abraded and bathed in pus, or if we have to judge of the leucorrhœa wholly from the discharge, and we find it composed of pus, mixed with blood globules, we may assume the condition present as the one above described.

In such cases, the appropriate remedies are *nitric acid, sabina, sepia, calcarea, copaiva, terebinth, trillium, mercurius, arsenicum*, etc.

It is utterly impossible for us, in the present condition of our Materia Medica, to construct a reliable repertory. Jahr mentions many remedies for "bloody" leucorrhœa, many of which are also mentioned under the head of "flesh-colored" and "reddish discharges." But how are we to decide whether the blood-globules come from the abraded cervix or vagina, or the interior of the uterus; whether it was really a bloody leucorrhœa, or a simple mucous leucorrhœa, mixed with the blood of a uterine hæmorrhage? The appropriate local treatment of this form of vaginal leucorrhœa with abrasion, consists of injections of *calendula, hamamelis,*

* West on Diseases of Children.

trillium, borax, sanguinaria, podophyllum; or *nitric acid, per-manganate of potash, bichromate of potash,* or *nitrate of silver.* Those first mentioned being mildest, may be tried first. In many cases they prove effectual. The proportion best adapted to the purpose is a watery infusion,* but the mother tincture can be used in the same proportion. The latter, unless largely diluted, cannot be used with safety during pregnancy. They are not as efficient when used weak, unless the case be a recent one. The *nitric acid* enema may be prepared ten or twenty drops to the pint; the *per-manganate of potash,* ten grains to the pint; the *bichromate* one or two grains to the pint, of water; the *argentum nitratis,* in the same proportion. In non-pregnant women, and long standing cases, stronger proportions can be used.

Gonorrhœa.—If this disorder has been contracted previous to the pregnancy, and its results—namely, ulceration, induration, and endo-metritis, are disposed to act as exciting causes of abortion, the *treatment* to be adopted should be the same as laid down for those affections.

If, however, the gonorrhœal infection occurs during pregnancy, an abortion may be caused by the congestion, irritation, and sympathetic fever which accompanies the outset of the disease. In such cases, the woman should be ordered to keep as much as possible the recumbent posture, use cooling and mucilaginous beverages, such as flax-seed, marsh mallows, or ulmus fulva, and one of the following medicines be prescribed; *aconite, gelseminum,* or *veratrum viride,* if the fever assumes a high grade, with local congestion; *cannabis, cantharis,* or *senecio aureus,* if the febrile irritation is accompanied

---

* One drachm of the crude material to one pint of water.

with much local pain and irritation; *sabina, terebinth,* and *pulsatilla,* if the irritation seems likely to bring on uterine pain and flooding. If this first stage passes over, and the second, or that mentioned on a previous page—namely, an inflamed and abraded surface or patch on the os uteri—obtains, the same treatment as that advised for "granular ulcer" should be adopted.

SECTION V.

ULCERATION OF THE OS AND CERVIX UTERI.

The *therapeutics* of ulceration of the os and cervix uteri cannot be considered at all settled. Those remedial agents which are appropriate in this disease, are those which are Homœopathic *to the pathological condition, as well as to the symptoms present.* Those who still assert that this disease can be diagnosed and treated from the symptoms alone, should ponder well over the statement of Professor Simpson, a statement concurred in by all the best and most practical observers of both schools of medicine—namely: "There can be no doubt of the fact, that there seems to be no organ in which there is a less strict relation observable between the intensity and character of the existing pathological disease, and the intensity and character of the accompanying symptoms, or between the exact nature of the structural lesions that are present, and the exact combination and succession of functional derangements to which they give rise." Dr. Madden[*]

[*] Uterine Diseases, p. 19.

says: "In this disease I have met with the greatest possible variety in the symptoms complained of, when a physical examination revealed the greatest possible correspondence between the pathological conditions of the various cases. I have found enlargement and ulceration of the cervix uteri in a case examined for another purpose, and when there was no detectable aberration from robust health. I have found the slightest form of cervico-metritis where the constitutional disturbance was very grave, and the symptoms had continued a long time; and lastly, I have known the symptoms to connect themselves so completely with some distant organ, as the head, the stomach, the heart, the mammæ, or the extremities, that indirect causes alone led me to suspect the uterine complication."

How is the mere symptomotologist to select the appropriate remedy for a case of ulcerated *os*, if he ignores all pathological conditions? He has nothing to do with *conditions*, therefore he must not avail himself of the general state of the patient—*i. e.*, whether she is disposed to ulcerations; nor can he be guided by the *discharges*, for a discharge is a pathological symptom. Even if we allow him to be guided by the appearance of the discharge, *per vaginam*, he would be as likely to be misled, unless the microscope revealed to him its *real* character.

It is because our Materia Medica is so incomplete, especially in objective symptoms (structural lesions), that our therapeutics of ulceration of the uterus are so uncertain and unsettled. Says Dr. Madden: "On going over the whole of the remedies which have been proved by Hahnemann and his followers, * * * I can only find four purely pathological symptoms which indicate organic changes in the uterus, and these are as follows:

(1) "Irregularity of the os uteri"—*natrum muriaticum*.

(2) "Metritis"—*secale cornutum*.

(3) "Softness of uterus"—*opium*.

(4) "Swelling of cervix"—*cantharis*.

It is needless to add that the symptoms noted under *natrum muriaticum*, and that under *opium*, cannot be trustworthy. *Secale*, in poisonous doses, causes a severe form of metritis, and the same may be predicated of *cantharis*. So that, of all the medicines in our Materia Medica, but two have a recorded and reliable objective symptom relating to the uterus.*

The condition of the knowledge of uterine pathology, at the time Hahnemann and his immediate successors made their provings, is a valid reason for the absence of such symptoms; but the *provers* of the present day cannot have any such excuse for not presenting us with pathological symptoms, if not on women, certainly upon female animals. Imagine a proving of *sepia, pulsatilla, sabina, sulphur, phytolacca, kali bichromat.*, etc., conducted in such a manner as to elicit subjective and objective symptoms. Then, indeed, could we prescribe with some degree of confidence for the disease under consideration.

For particular information concerning the Homœopathic treatment of ulceration of the os and cervix, I would refer the physician to "Madden on Uterine Diseases," "Jahr on Diseases of Women," "Gollmann on Diseases of Urinary and Sexual Organs," and the papers which have appeared in our various journals,

---

* *Aquaphobin* causes "inflammation of the uterus in cows." See "Proving of *Aquaphobin*,"—*Phila. Jour. of Hom.*, vol. iii., p. 262.

also Dr. Ludlam's forthcoming work on Diseases of Women.

Madden found the most useful remedies in uterine diseases generally, to be *pulsatilla, sepia, sulphur, nux vomica,* and *platina.* He also used local applications of *nitrate of silver, per-nitrate of mercury,* and *potassa fusa,* and claims that these drugs have a specific action on uterine disorders, acting Homœopathically, whether as constitutional or local remedies.

Jahr recommends (1) *nitric acid* and *thuja;* (2) *arsenicum, belladonna, china, cicuta, cocculus, mercurius, pulsatilla, sepia, silicea,* and *sulphur.* Of these, *nitric acid, arsenicum, mercurius, sepia, silicea,* and *sulphur,* are most worthy to be relied upon. It is doubtful if the others are Homœopathic to ulceration at all. Jahr mentions a few others which are probably appropriate, namely, *calcarea, aurum, graphites, petroleum, ruta, sabina, secale, hepar sulph.,* and *lachesis.* Dr. Marcy[*] recommends against "simple inflammation of the mucous membranes lining the cervical canal and the cervix uteri," *sepia, secale, sabina, pulsatilla, cocculus, apis mel., calc. carb.;* also, *mezereum, conium, nitric acid, petroleum, bovista, borax, platina, kali carb., ferr. iod.,* and *sulphur.*

For "*suspected* ulcerations," he has found most useful, *secale, apis,* and *thuja;* also the above mentioned remedies.

Leadam seems to teach that "simple **induration**" of the os uteri and cervix uteri, precedes ulceration, and is the result of *chronic* inflammation. The remedies which he advises for this condition are, *belladonna, platina, calendula, sepia, conium, arsenicum, iodium, aurum,* etc.

---

[*] N. A. Jour. of Hom., vol. v., p. 80.

The constitutional remedies which observation, experience, and information gleaned from all sources, indicate as most appropriate for the *inflammation* and engorgement, which precedes ulceration, are—*sabina, secale, cimicifuga, caulophyllin, belladonna, pulsatilla, apis mel., aquaphobin, podophyllum, murex, mitchella, cantharis,* and *platinum chlor.,* when the condition is acute or recent. In chronic cases, *sepia, sulphur, helonias, aletris, platina, lachesis, conium, thuja, and merc. iod.,* are more often indicated.

Any remedy which has been found useful, or is indicated by its pathogenesis, in inflammation and induration of other portions of the body having a like tissue with the cervix uteri, will be likely to prove serviceable in this affection.

In the treatment of inflammatory congestion, and induration of the os and cervix, the use of topical applications should not be neglected. The Homœopathic school are divided in opinion, as to the propriety of such applications, the *conservative*, or Hahnemannian party, claiming that the internal administration of the proper remedy is alone sufficient to bring about a cure—the *progressive* party believing it necessary and proper to apply the *same* medicine which is prescribed internally, to the diseased portion or tissue, or medicines that by their local action are Homœopathic to the lesion present, when *topically* applied.

I incline to the latter belief, because my observation and experience has shown me that our success is greater when remedies are applied to the diseased structures, topically, as well as when reaching them through the medium of the general circulation.

The *treatment* laid down in these pages is consequently based on this belief. At the same time, I am willing to allow the largest liberty of action to all the members of the Homœopathic school.

The *topical* treatment of the inflammatory condition which precedes ulceration, should consist of the application by enema or lint, of the following medicines (or others which seem indicated).

(1) For simple inflammation—*aconite, gelseminum, arnica, calendula, baptisia, cimicifuga,* and *secale.*

(2) Inflammation with induration—the above, and also *urtica urens*, **belladonna**, *kali hyd., kali brom., conium,* and *clematis.*

These remedies should be prepared in proportion of ten or twenty drops of the mother tincture to four ounces of water, and the whole quantity used at once; or if the remedy is applied by means of *lint*, it should be saturated with a preparation made in proportion of ten or twenty drops to one half ounce of water. The water may be cool or warm, as the physician judges best. It should be remembered that cold applications to the os uteri, during pregnancy, are sometimes capable of exciting reflex irritation.

The *syringe* used for such purposes may be the common vaginal syringe, made of hard rubber, and holding two ounces; or one of the various instruments which are sold. I prefer the Essex syringe, with the air chamber. The tube* should be inserted so that the bulb is *nearly* in contact with the os, and the fluid should be thrown in a *slow, continuous* stream. If thrown with too much force, uterine contractions (in case of pregnancy) may be caused. If lint is used, it should be applied through a speculum; it should be pressed gently against the os,

---

* A *large* vaginal tube, with a *large* bulb, perforated with holes one-sixteenth or one-twelfth of an inch in diameter—should be obtained with the syringe. It is made expressly for this use, and can be purchased from Mr. Halsey, the publisher of this work.

and kept in *situ* with a sound, or anything which will answer that purpose, while the speculum is withdrawn. The best time to apply medicated lint is at night; it may be removed in the morning; this the patient can generally do; indeed she can often apply it herself, if the uterus is low in the pelvis. These general directions are applicable in case of any disease of os and cervix, when remedies are to be used topically.

We will now consider briefly the *treatment* of the varieties of ulceration.

Simple Granulating Ulcer.

The remedies most generally useful are *arsenicum, platinum chlor., phytolacca, stillingia, sabina, kali hydriodatus, mercurius iodatus, secale, sepia, nitric acid, podophyllum, hepar sulph., calendula, kreosotum, nymphæa odorata, cornus circinata, kali chlor., hydrastis canadensis, argentum nitricum.*

The same remedies should be applied topically in such proportions as have been heretofore indicated in this work, leaving the special preparations (as *nitrate of silver*, etc.) to the judgment of the practitioner. My experience with *arg. nit.*, in simple erosions, leads me to prefer the enema to be very weak—2 grs. to 8 ℥— and frequently used.

Varicose Ulceration.

The practical experience of physicians, and the pathogenesis of medicines, seem to indicate that this variety is best treated by use of

Collinsonia, hamamelis, lycopodium, pulsatilla, graphites, lachesis, thuja, sepia, trillium, sulphur, nux vomica, and *arnica,* prescribed for internal and topical use.

The *hamamelis* is the most generally useful remedy in this disease. But to be efficacious it should be used in material doses (10 or 15 drops of the tincture internally) and the watery infusion* used as an enema. *Thuja, trillium, nux vomica,* and *arnica,* may be also used in infusion, but of less strength.† *Lycopodium, lachesis, graphites,* and *sulphur,* are efficacious internally, in the 30th or 200th potency.

Fissured Ulcer.

In the Repertories we find recommended for rhagades generally, *alumina, calc. carb., hepar sulph., lycopodium, mercurius, petroleum, rhus tox.,* and *sulphur*.

It is probable these may be equally useful in fissured ulcer of the os uteri. But in my estimation the chief remedy is *nitric acid*. Those who are acquainted with the specific curative power of this remedy in fissures of the anus, fissures of the tongue and lips, can readily believe this assertion. It has cured this lesion, in the low and high dilutions, with or without its local application, yet I am inclined to recommend that, in old and severe cases, it be applied locally. This may be accomplished by the use of an enema in the proportion of ten drops of the dilute acid to four ounces of water; or the dilute acid itself, applied with a soft brush. The ulcer should first be wiped *dry to the bottom of the fissure*, with a piece of lint attached to a probe, and the brush pushed into the depth of the ulcer. The remedies first named, also *nitrate of silver*‡ and *glycerole of aloes,* may be applied in a similar manner. The *glycerole of aloes* is already widely and favorably known as almost

---

* *Hamamelis* (the bark) ℥ i—warm water 1 qt.—infuse 3 hours.
† ʒ i of the powder, leaves, or flowers, to 1 qt. water.
‡ *Arg. nit.* (crystals) gr. x.—*aqua dest.* 1 ℥.

a specific for rhagades in the skin. In the few cases in which I have used it in similar states of the mucous membranes, it has proved equally efficacious. The *bichromate of potash* has been found very useful in certain conditions similar to fissured ulceration. Dr. Dudgeon used it internally and topically.

Follicular Ulceration.

The remedies are (1) *merc. iod.* and *bin-iod., kali iod., kali brom., phytolacca, stillingia, baptisia, sabina, sulphur, nitric acid, silicea, sanguinaria, thuja.*

(2) *Argentum nit., per-manganate of potash, chlorate of potash,* and *potassa fusa.*

The *first* class should be used internally and topically; the *second,* in most cases topically alone, although in some cases their internal administration may be appropriate.

If the follicular ulcer is seen in its incipient stage, one or two applications of the crude *nitrate of silver* to the surface will suffice to heal it: at a later period, when an *excavated* ulcer is present, the strong solution of that caustic, or the *per-manganate of potash* (of the same strength) should be applied with a pointed brush, pushed to the bottom of the excavation. In some cases, according to Madden and Tilt, the *potassa cum calce,* applied in the form of the crude *stick,* causes rapid filling up of the ulcer.

Phagedænic Ulcer.

This is the "corroding ulcer" of Leadam, who is sanguine in the recommendation of *arsenicum* as the principal remedy. That author also advises *sulphur, secale, lachesis, iodium,* carb. veg., and *pulsatilla.*

With the single exception, however, of *lachesis*, it is extremely doubtful if any of the last-named are Homœopathic to phagedæna, in any form: not even *secale*, which is only useful in *dry* gangrene, if in gangrene at all.

The remedies appropriate in this form of disease — true phagedænic ulceration (not cancerous), are, *nitric acid, muriatic acid, per-manganate of potash, hydrastis canad., phytolacca, iodide of arsenic, arseniate of iron, tartrate of iron and potassa; nymphæa odorata, merc. bin-iod., kali hyd.* and *stillingia*.

These remedies should be used thoroughly and perseveringly, internally and topically, as before described. The *tartrate of iron and potassa* was said by Ricord to be the "sworn enemy of phagedæna." One patient, with an ill-looking ulcer of this character upon the os, I cured with the first trituration of this remedy alone. There was present considerable anæmia and debility. Dr. P. H. Hale, of Michigan, uses the *nymphæa odorata* (white pond lily) with much success in this form of ulceration.

The remedies, however, which have been most useful in my hands, for phagedæna, whether syphilitic or not, have been the *bin-iodide of mercury*, in alternation with *kali hyd.* It is a curious fact, but one observed by others besides myself, that some remedies will cure promptly when given in alternation, after they have failed singly. This is the case with the above remedies. Neither the *bin-iodide* nor *iodide of potash* were capable of curing alone, but when alternated, they removed the diseased condition rapidly. I usually prescribe the former in the third decimal trituration: the latter in solution, after the formula given below* (about the

---

* ℞ *Kali hydriodatus* ℈ i.
    Aqua dest. or *syrup. simp.* ℥ viii. One teaspoonful three times daily.

second dilution): each three times daily: the former before meals, the latter about two hours after. I know of no treatment so uniformly successful, and strongly advise its adoption. Dr. W. T. Helmuth informs me that he uses successfully, in this condition, the same medicines in nearly the same manner.

Next to this, I prefer the use of *per-manganate of potash*. *Manganese* is a blood-restorer, like *iron*, and the large amount of *chlorine* contained in the preparation, makes it a powerful remedy over diseased states which lead to phagedæna. Besides, it is a powerful but safe caustic, acting as a local Homœopathic remedy. The third dilution may be administered internally, and the strong solution (ten grs. to one ounce of water) applied with a brush or lint.

Syphilitic Ulceration.

The treatment of chancres on the os uteri, or the ulceration which follows it, does not materially differ from that adopted for the last named variety.

Simple chancre of the os will often heal kindly under the use of *mercurius sol.*, or *merc. iodatus*, or *nitric acid*, together with a wash of *aqua calendulæ*, dilute *nitric acid*, or the preparations previously mentioned for enemas.

Indurated chancre requires the same mercurial remedies, aided by *kali hydriodatus*, *stillingia*, or *phytolacca*.

Phagedænic ulcerations, of a syphilitic character, require the same treatment as phagedæna arising from non-specific causes. In this affection I have used the *merc. bin-iod.* and *kali hydriodatus* in alternation, with the happiest results.

Aurum, kali bichrom., platinum chlor., iris versicolor, and *arsenicum,* have all been highly recommended in syphilitic ulcers, but my experience with them has been so limited, that I am unable to report favorably in relation to their therapeutic value.

Resumé.

There are many remedies not specially indicated in ulceration of the os and cervix, which may be useful and indispensable to the cure of that affection.

It is well known that certain states of the system, namely, anæmia, psora, some miasmatic or dyscratic poison, may retard and even prevent the cure of local or general diseases of a different character. Thus, if considerable anæmia be present, we may select the remedy for the local affection with ever so much care, but if we do not take into account the general condition, and select some remedy to meet it, we shall fail to effect a cure. It is very rarely the case that any one remedy will meet both conditions, namely—the ulcer (syphilitic or not) and the anæmia; we therefore are obliged to alternate two remedies, for example, *ferrum* and *mercurius: china* and *phytolacca: hydrastis* and *kali hydriodatus.*

Again, if a psoric taint exist in the organism, we will have to act upon the advice of Hahnemann, and give an occasional dose of some anti-psoric—*sulphur, calcarea,* etc. (30th or 200th), to antidote that miasm, and the cure will afterwards progress rapidly.

In miasmatic districts we sometimes have to use *arsenic,* or *quinine,* before other remedies will exercise a curative effect on the disease.

During the prevalence of epidemics of *diphtheria,*

ulcers of all kinds are apt to take on a diphtheritic character—ulcers situated on the os and cervix are not an exception. When this occurs—and the careful physician will be on the alert for such a complication—the remedies recommended for diphtheria will have to be resorted to.

I hardly need add, what every practical physician should know, that during **the treatment of** ulceration, the physiological functions of the various organs should be kept, as much as possible, in a normal condition. Respiration, digestion, and depuration, should go on properly, or the cure will be retarded.

The *diet* of the patient should be varied to suit the exigencies of each individual case, the use of stimulants advised or not, as the system demands. The amount and character of *exercise* should **be regulated in the** same manner.

It has always appeared to me that the *postural* treatment of ulcers of the cervix, has not received the attention which it demands. If we are treating an ulcer on the foot or leg, we know it will heal in much less time if the limb is placed horizontally, so that the column of blood does not press upon the irritable tissues in or near the ulcer. Should not the same hold **true, to a certain degree, in the case of** ulcers situated upon the most depending portion of the uterus? We are aware that some writers do advise the patient to assume and remain in the recumbent position while under treatment, but not for the reason above alluded to. I am satisfied that if this suggestion be borne in mind by the physician, and acted upon in certain cases, his success will be much greater, especially if he is treating **a case of** ulceration occurring during pregnancy, when, more than at any other period, the uterus is loaded with blood.

There is another reason why the recumbent posture should be advised. It is well known that ulcers and abrasions of the os, are kept open, and irritated by the constant contact with, and rubbing against the posterior wall of the vagina, upon which the cervix rests in most cases in the early months of pregnancy. The most carefully selected remedies will not effect a cure, when this irritation is kept up. In non-pregnant women we can insert a pessary, and lift the uterus up from the vaginal wall, but during pregnancy the use of pessaries are generally objectionable, and our best means of aiding the medicinal treatment is to allow the patient to stand or sit but a small portion of the time, or until the ulcer shall be covered by a healthy mucous membrane.

It is in these cases, especially in women of large size, that the abdominal bandage is found a useful mechanical auxiliary. The best bandage is made of elastic silk and rubber, woven whole, or laced at the back or sides.

On account of the irritation and congestion consequent upon the act of *coition*, it is best to prohibit sexual intercourse almost entirely during the treatment of uterine ulceration in pregnant women.

Ovarian Diseases.

When we have to deal with a *simple ovarian irritation* from perverted physiological influences, the treatment will be nearly the same as recommended for "Return of Menstrual Crisis,"—namely, the avoidance of all sensual emotions, coition, etc., and the use of remedies which have a sedative effect upon the ovaries. These are (1) *cantharis, cannabis indica, apis mel., platina, sabina, podophyllum, lachesis, zincum valerianatum,* all of which should be used in the higher dilutions; and (2) *bromide of potash,* and *conium,* if the remedies of the first class fail to relieve. The latter medicines must, however, be given in doses of one grain, or more, of the $\frac{1}{15}$ trituration, twice or thrice daily.

Neuralgia of the ovaries (if such a disease exists) requires the use of *atropine, aconite, apis, colocynth, lachesis,* and *zincum valerianatum.*

Congestion and inflammation is treated successfully with *aconite, veratrum viride, gelseminum, belladonna, mercurius, apis mel., lachesis,* and *platinum.* Ovarian *tumors* and dropsy may be held in check by *apis, lachesis, lycopodium, arsenic, iodine, kali brom., kali chlor.* and *phytolacca.* In some cases the ovaries can be punctured, and the escape of pus or serum may allow pregnancy to go on undisturbed. It is quite doubtful if a diseased ovary could be removed without resulting in abortion; cases might occur where the experiment would be justifiable.

SECTION VI.

UTERINE DISPLACEMENTS.

The *treatment* of displacements of the uterus during pregnancy in such a manner as to avoid the risk of an abortion, is a matter of considerable difficulty. Most writers on Diseases of Women insist upon the employment of medicinal and postural measures alone, and assert it to be improper to resort to mechanical appliances in any case. But the *absolute* banishment of

Prolapsus Uteri.—A certain amount of prolapsus is a common accompaniment of pregnancy in the early months; but after a time the uterus no longer descends, the ovum acting as a very efficient intra-uterine pessary. If this does not occur at a proper time, all that is required is a return of the uterus to its normal position by gentle pressure, and the continuance in the recumbent position for a time, together with the administration of the following remedies:

Belladonna, if the pelvic organs are abnormally congested, with throbbing at the cervix and heat of the parts, observable by the touch.

Nux vomica, when the vaginal and other muscles upon which the uterus depends for its support, are in a relaxed, atonic condition, and there is constipation from torpor of the bowels. This remedy may be used also as a vaginal enema.

Podophyllum, collinsonia and *æsculus*, are applicable if there is general fullness of the veins of the pelvis—especially the hæmorrhoidal veins—constipation or diarrhœa, and uterine tenderness.

Caulophyllin, macrotin and *secale* if the pain and pressing-down in the uterus are the most prominent symptoms.

Pessaries are not generally well borne, owing to their tendency to excite reflex irritation. In two cases, however, of prolapsus from great relaxation, I used the *ring*-pessary for nearly two weeks, without any other than beneficial effects. These patients were quite debilitated, but this was overcome by the use of *citrate of iron and strychnia* and *helonias*, with a generous diet.

If adhesions have taken place, abortion is inevitable, and may as well be hastened, and the patient saved the risk of a miscarriage at a later period.

Anteversion.—In the first month or two of pregnancy a slight anteversion of the uterus is its normal position. If this is greater than natural, an elastic abdominal bandage should be worn, the patient advised to lie a good deal on her back, and the rectum emptied daily by an enema. The medicines most likely to be of use are *belladonna, nux vomica, cimicifuga, caulophyllum, podophyllum, senecio gracilis, and collinsonia.* For the irritation of the bladder we may give *cantharis, cannabis, chimaphilla, senecio aureus, sabina, terebinth;* and use the catheter if necessary.

Retroversion.—The treatment of retroversion of the pregnant womb should be conducted with great caution, unless we are satisfied that adhesions exist, or such an amount of impaction is present as to utterly preclude its replacement, in which case the induction of abortion is an absolute necessity, unless nature does not perform the operation herself, and relieve the uterus of its contents. When we have ascertained that a *retroflexion* exists, every means in our power should be adopted to prevent further displacement. The bladder should be kept empty, and the bowels never suffered to go more than a day without evacuation. For this latter purpose, the use of *nux vomica, podophyllum,* and *sulphur*, should be aided by aperient food — coarse bread, fruit, etc.—and, if necessary, the use of injections.

If, however, retroversion has occurred, attempts to restore it to position should at once be adopted. The rectum and bladder must be emptied of their contents, and the fundus-uteri be elevated by gentle, firm and persistent pressure by the two fore-fingers of the right hand. If this does not succeed, the patient should be placed on her knees in bed, with her head lowered; the fore-finger of one hand should be introduced into the vagina, the other into the rectum; the os should then be drawn down with the one, and the fundus elevated with the other. In some cases the finger is unable to seize the os, so as to exert the necessary traction, in which case the bent extremity of the uterine sound may be carefully introduced a short distance into the os, and the uterus drawn downward thereby.

M. Gariel has proposed that one of his vulcanized india-rubber pessaries should be introduced into the rectum, and that the fundus-uteri should be raised by inflating the pessary. I do not know if this plan has been tried in practice, but it would probably be effective in a case admitting of mechanical re-adjustment of the uterus. If it should cause abortion, it would only hasten an inevitable result. Tyler Smith thinks it might lacerate the soft parts unless great caution was observed. The instrument itself consists of a dilatable air pessary, terminating in a tube, and an air reservoir, with small taps affixed to each. After immersion in warm water, the collapsed pessary is insinuated into the rectum, behind the uterus, by means of a probe. The air reservoir is then fitted to the tube of the pessary, the taps are opened, and by the pressure of the hand the air contained in the reservoir is transferred to the pessary, so as to lift the uterus out of the hollow of the sacrum. This instrument is also called a *colpeurynteur*, and is sold at all Pharmacies.

The uterine sound should never be used if the continuance of pregnancy is desired, or unless previous abortions during gestation, owing to the inability of replacing the uterus, should remove all hope of conducting the pregnancy to a natural termination.

The "*levator perinei*," invented by Dr. Sims, for the purpose of exposing the whole of the vaginal walls as well as the os uteri, has been used by Dr. Helmuth with success in a case of retroversion during pregnancy. When the necessary traction was made, the uterus quickly resumed its normal position. To use this instrument, place the patient as nearly as possible lying upon her breast and stomach, her left arm thrown behind, and the chest rotated forward, bringing the sternum quite in contact with the bed; the feet drawn up, one extremity of the instrument is to be inserted into the vulva, and by the other the perineum is to be forcibly lifted up, allowing the atmosphere to dilate the vagina. This will give a full view of the entire vaginal cavity— "better," says Dr. Gardner, "than the ordinary cone-shaped, or even the many-valved instruments." Thus, besides its use of replacing the retroverted womb, this contrivance may be used instead of the speculum for the diagnosis, and as an aid in the treatment of the various affections of the os cervix, and vagina.

For the *special* medicinal treatment of retroverted uterus, reference is made to my treatise on that subject.*

The principal remedies to be used, after replacement by mechanical means, and the enjoinment of a position on the face or side when in bed or in the recumbent position, are: *aletris, helonias, nux vomica, sepia, ignatia, macrotin, iodide of iron, podophyllum,* and *secale*.

In plethoric women, or those with a large abdominal development, the use of an elastic bandage, worn while the patient is on her feet or sitting up, is quite an aid to our efforts in the prevention and cure of this

---

* Therapeutics of Retroversion of the Uterus.

malposition. It acts by lifting the weighty intestines off from the uterus, heavier than normal with congestion or its natural contents, and it also allows the relaxed muscles which support the uterus to regain their strength and tone.

SECTION VII.

REMEDIAL TREATMENT OF ABORTION.

The *remedial* treatment of abortion may be divided into

(1) *The Mechanical.*
(2) *The Medicinal.*

The *mechanical* consists of those measures which are to be used when the symptoms of abortion have already set in, and may have in view two ends—namely, (*a*) the arrest of the morbid process, and the saving of the life of both mother and child, and (*b*) the expulsion of the fœtus and placenta, and the safety of the mother.

The *medicinal* consists in the use of those remedial agents which have for their object the same ends as the above; also the palliation of the painful and dangerous symptoms which arise during the progress of an abortion.

These cannot be treated of separately, as it would cause confusion and needless repetition, we shall therefore allude to both in such places as are necessary.

I here feel it my duty to protest against the stereotyped recommendations of Homœopathic writers relative to the treatment of some of the early symptoms of abortion.

We are told that if the accident is caused by a fall, blow, or concussion, we must give *arnica*, and if from a strain, we should prescribe *rhus tox*. Why? Because the **former is a good remedy for the effects of a fall, bruise, or concussion, and the latter those following a "*strain*."** This is as absurd as it is unscientific. Yet those who have **recommended this treatment are the greatest sticklers for** *symptomatic* indications.

The fact is, *arnica* is by no means a specific for the first symptoms of abortion, any more than any other remedy used by us in similar cases. The mere postural treatment which the patient is subjected to, after a fall, will sometimes succeed in **warding off** the abortion, while, without such treatment, *arnica* would be of no benefit. Regarding *rhus tox.*, it is **very doubtful if that remedy is ever** indicated in abortion.

From whatever cause the symptoms of an impending abortion may arise, the remedy must be selected strictly in accordance with the *symptoms* of the patient, and the pathological condition known or supposed to exist, and in addition such postural and dietetic rules as seem appropriate.

In nearly all cases, as soon as any of the symptoms mentioned above, as indicating an impending abortion, appear, we should insist that the patient immediately divest herself of all tight and heavy clothing, and assume the recumbent position on a bed or lounge, and lie as still as possible, avoiding all sudden movements, emotional excitement, or any effort whatever involving the motion of the abdominal viscera.

Her diet should be of the simplest character, very light, easily digestible food, cool beverages, and the avoidance of all warm and stimulating drink* or food. It is sometimes well to elevate the pelvis somewhat higher than a level position, by placing a folded blanket under the hips, especially where there is present congestion of the uterus, or a previous concussion; also in cases of prolapsus. In retroversion, the position should be on the side or face.

If the symptoms consist of the usual *pains*, unattended by discharge of blood, one of the following remedies will be most likely to remove the uterine irritation, and arrest the abortion: *caulophyllum, cimicifuga, aletris far., chamomilla, secale, gelseminum, nux vomica, belladonna, tanacetum*, or any medicine capable of arousing the motor action of the uterus.

If, on the contrary, we have hæmorrhage, without any or but little uterine pain, we shall find most useful—*sabina, cinnamon, erigeron, ipecacuanha, secale, sulphuric acid*, and *arnica*.

If both hæmorrhage and pain are present, the remedy which will cover the whole group of symptoms must be selected, or failing in this, two remedies may be alternated at such intervals as shall seem proper.

I cannot advise the use of applications of cold or warm water in such cases. The danger of reflex action from them more than outweighs any benefit which may accrue.

[There is one remedy much relied upon by the Allopathic school, in the treatment of threatened abortion, which is certainly very successful, if a large amount of testimony is sufficient to prove its usefulness. This is the administration of full doses of *opium, laudanum*,

---

* I have known an exception to this. In several cases of threatened abortion at the second month, when pain and flowing had already set in, the symptoms were permanently arrested by a wine glass of *hot gin and water*. Did the juniper act as a Homœopathic remedy in these cases?

or *morphine* (the former are said to be most efficient), as soon as possible after the appearance of the symptoms. The dose of *laudanum* usually prescribed, is from twenty to forty drops.]

Should the means above recommended fail to arrest the premonitory symptoms, and those appear which indicate a rupture of the ovum; or extravasation of blood between the membranes and the uterus; or the separation of the placenta—namely, chills, with more or less regular pains and hæmorrhage—we may still administer remedies in the hope of preventing a termination in actual miscarriage. I consider the probabilities of such favorable termination as very small when the above symptoms have set in, but the following remedies may be tried:

Aconite, if there is chill, with shivering (rigors), with anxiety, coldness of the extremities; sensation of heaviness in the uterine region.

Belladonna, for pressing and tensive pains in the whole abdomen, with sense of constriction or distension; lumbar pains, as if the sacrum would break; pressure downward in the abdomen and pelvic organs, as if everything would fall out.

Caulophyllum, when the pains are regular, like labor pains, the os uteri relaxed, some discharge of bloody mucus, and sometimes cramps in the extremities. This remedy is particularly useful in cases of habitual abortion from any cause. It should be administered daily for a week before the period of the usual abortion, or as soon as any suspicious symptoms appear. It will often prevent or arrest a premature labor in the last months of pregnancy.

Cimicifuga is useful in cases where the pains are similar to those under *caulophyllum*, but attended with intense headache, as if the brain and eyes were pressing from within outwards; soreness of the uterine region, relaxation of the os and vagina. It is indicated for women subject to rheumatism, or spinal irritation, and hypochondriacal lowness of spirits.

Gelseminum, when there is general prostration and rigor, without coldness, with paralytic sensations in the extremities, fullness and dullness of the head, and obscuration of vision. The pain and hæmorrhage are not very marked. It is indicated in women subject to depressing emotions, and in cases where the symptoms were apparently brought on by fright or fear.

Cinnamon, if the hæmorrhage is profuse and bright red, without other notable symptoms. In similar cases *erigeron* and *erechthites* are indicated.

Ruta graveolens, when there is great prostration, confusion of mind (sometimes pain in the stomach and violent retching), double vision, feeble pulse, cold extremities, twitching in the limbs, and intense pain in the sacral region.

Sabina for "discharge of dark-colored coagulated blood, pressing and drawing pains, from the small of the back to the sexual parts; soft and flat abdomen; continual urging to stool, with diarrhœa, nausea and vomiting; fever, with shuddering and heat." Also for cases accompanied with inflammation of the small intestines (enteritis and peritonitis), jaundice, and excessive irritation of the urinary organs. It is most applicable to plethoric women, whose menses are habitually profuse and painful.

Secale is most useful in cases where the uterine pain is *constant* and *unintermitting*, and in which there is organic disease: or deficient vitality of the uterus with hæmorrhage of black and thin blood, fear of death, pulse small and most extinguished, os open and dilated. It is suitable to feeble, exhausted, cachectic women with disposition to passive hæmorrhage or convulsions.

Pulsatilla for intermittent hæmorrhage, recurring every now and then with redoubled violence, accompanied with expulsive pains and discharge of dark blood with coagula.

Many other remedies might be enumerated, especially those mentioned under the head of "*Medicinal Causes*," each of which may in certain cases be indicated.

Eclectic physicians have great confidence in *aletris farinosa*, asserting that it will arrest abortion, even after severe hæmorrhage has set in.

Allopathic treatment is generally worse than useless. It may be summed up thus: bleeding, leeches, cold applications, astringents internally and externally, dry cupping and drugs. There is one remedy, however, which I have before alluded to, in use by that school, which seems to be successfully used, namely—*laudanum*.

"This remedy," says Cazeaux, "is one of the most efficacious, and sometimes it alone has enabled us to arrest a labor, whose termination seemed to be inevitable, and thus has permitted gestation to pursue its regular course."

"I cannot refrain from citing the following instance in illustration: A woman advanced to three months and a half, was taken with pains in the abdomen and loins, after a violent altercation with her husband; on the following day the pains augmented and a little bloody fluid escaped from the genital organs; the pains still continuing, and the discharge being somewhat increased, on the third day the patient came on foot to the Clinique. I found, on her arrival, that the uterine contraction was very distinct, the pains sharp, and renewed every eight or ten minutes; pure blood was discharging from the vulva, and the orifice was sufficiently dilated to *permit the finger to pass readily up as far as the naked membranes.* I administered sixty drops of *laudanum,* divided into three doses, which were given at intervals of three-quarters of an hour, and by the end of this time, the pain disappeared, everything resumed its natural order, and the gestation went on till full term."*

Cazeaux says he might multiply such cases almost *ad infinitum.* Other obstetric writers give similar testimony in favor of *opium.* In view of this we should not hesitate to use it in cases of abortion when the usual remedies seemed powerless. It is not strictly Homœopathic, but should be considered as a dynamico-mechanical remedy, in the same class with *chloroform, ether,* splints and mechanical appliances generally. It acts by producing perfect sedation of the uterus. It is asserted by some physicians, that large doses of *gelseminum* will have the same effect, but it is not so manageable and reliable as *laudanum.*

Sulphuric acid, in quantities as large as can be safely borne, is said to have been used very successfully, but I have had no experience with it, except in cases of hæmorrhage after abortion.

These measures, which have for their object the saving of the life of the ovum, must not be persisted in too long. When the hæmorrhage is so profuse as to endanger the safety of the mother, our attention must be entirely turned in that direction.

The first object to be attained is, of course, the *arrest of the hæmorrhage.* We will suppose that all the medicines most likely to effect that object, and which we have had time to use, have been tried, *and the os uteri is not dilated or dilatable:* we must now resort to other means. The most important of these are, (*a*) THE

---

* Braithwaite, Part 46, page 208.

TAMPON; (*b*) COLD WATER AND ICE; (*c*) HOT WATER. The *tampon*, when well applied, acts in two ways: 1st, by opposing the escape of blood externally, thus forcing it to coagulate, and consequently to obliterate the bleeding vessels; 2d, by irritating the womb by its contact, thereby determinating its contraction, and the expulsion of the product of conception. This circumstance, indeed, is one of the best-founded objections to the use of the tampon in the early months of gestation. But it appears to me to be rather an advantage than otherwise. The cessation of the flooding is always a necessary consequence of the uterine contractions. The mother's life is not bought too dear, when it is saved by the expulsion of a fœtus. Cazeaux asserts that the use of the tampon is not always necessarily followed by abortion, but the exceptional cases must be rare. However, it matters not: the life of the mother is of the most importance.

The tampon should not be used after the sixth month: even at the fifth month, the physician should carefully watch the body of the uterus, after the tampon has been applied, to assure himself, every moment, that its volume is not increasing. There is but little reason to fear the conversion of an open into a concealed hæmorrhage before the fifth month. Unless there is abnormal relaxation of the uterine walls, it would seem impossible for a large quantity of blood to accumulate in that organ. There is, perhaps, the possible danger of an escape of blood through the Fallopian tubes into the peritoneal cavities; but as those tubes are blocked up during gestation, such a result would be highly improbable.

The introduction of the tampon is generally performed in the following manner: "Some dossils or pellets of charpie are prepared—sometimes dry, at others smeared with *cerate*—and the vagina is then stuffed with these gradually, care being taken to have the first portions applied directly to the uterine neck: it would be better, perhaps, to connect them by means of a thread, so that they can afterwards be withdrawn more easily. When the vagina is filled, some thick masses of charpie directly over the vulva to sustain the pellets, and the whole is held in position by a T bandage."* (See note, *Speculum*.) But it is not always that we have the time and opportunity to procure such choice material, and we must not hesitate to avail ourselves of anything which will answer the purpose. We can use soft linen or cotton cloth cut in strips or square pieces; a soft linen or silk handkerchief; an old napkin; or even a piece of sponge. Strips of cloth, napkins, etc., may be introduced thus: wet them in tepid water, and placing one end or corner against the end of the forefinger, push it up the vagina until it comes in contact with the mouth of the womb; withdraw the finger carefully and push the rest of the cloth slowly, carefully, and by degrees, into the vagina until the canal is compactly and closely filled.

A *sponge* may be wet with cool water, or *hamamelis* water, and pushed up the vagina. The blood infiltrating into the pores of the sponge, soon coagulates and forms a voluminous clot, which seals up the vagina hermetically, without giving rise, says Dewees, to any of the accidents produced by the ordinary tampon: besides, it is borne without inconvenience, and may be left there until the expulsion of the ovum, although it would be better, says Cazeaux, to remove it after the lapse of a few hours, to observe the progress of the

* Cazeaux's Midwifery, page 345.

dilatation, and then replace it if the neck is still closed. I have never had much confidence in the sponge, for the reason that it is not as sure to prevent the escape of blood externally, as other materials.

The *colpeurynteur*, or dilatable india-rubber pessary, may be used as a tampon, filled with air, or better, with ice water. For directions how to use this instrument, see "Treatment of Retroversion," on previous page; also the *North American Journal of Homœopathy*, vol. 9, page 312.

In the few cases in which I used the colpeurynteur, I was not pleased with its use. I could not expand it sufficiently to prevent the escape of blood between it and the vaginal walls. Any tampon should be removed every six or twelve hours, and always within twenty-four, for the reasons—1st, to ascertain if the os uteri is sufficiently dilated; and 2d, to find if the fœtus, or placenta, or both, have been expelled. As has been stated previously, it is rare that serious hæmorrhage occurs before the expulsion of the fœtus, we shall generally be required to use the tampon after that occurrence. It was also stated that the danger from hæmorrhage usually ceased after the expulsion of the placenta (except after the sixth month), therefore, if upon removing the tampon we find the fœtus and placenta, or placenta alone, attached to the tampon, or lying in the vagina, we shall not need to insert the tampon again.

With the use of the tampon, we may advantageously administer such remedies as *caulophyllum, macrotin, gossipium, erigeron, secale* or *opium*, in material quantities, or sufficient to cause expulsive contraction.

Cold Water and Ice.—The application of cold water and ice to the epigastrium, thighs and uterus, to arrest hæmorrhage, is a very old practice. The theory is, that cold causes uterine contractions, and also decreases the calibre of the blood vessels. It is directed to be applied in several ways, namely: by means of compresses wetted in cool or ice water, and applied directly to the hypogastrium and vulva, or by a bandage which is worn around the thighs. It has been recommended to use ice water injections *per vaginam*, and even to plug the vagina with pieces of ice. Bags of pulverized ice have been placed upon the hypogastrium directly over the uterus. The applications may each be resorted to in cases where the exigency seems to demand. I prefer, in all cases prior to the fifth month, to use the tampon; but the materials for a tampon may possibly not be at hand, or the sensitiveness of the tissues may be so extreme that it cannot be used, in which instances cold water may be applied immediately. In several cases I have used *snow*, applied directly to the uterine region, or with only the thickness of a napkin intervening, After the fifth month, these applications of cold water, etc., are necessary adjuvants to medicinal remedies; but before that time, I do not consider them of such permanent utility as to warrant us in depending upon their use alone. In order to be of value they have to be continued for a long time, or until the placenta has been expelled, during which time the woman is exposed to injury by the constant contact of wet bed-clothes. Many of the disastrous sequelæ of abortion I believe to proceed from the use and abuse of cold water.

Hot Water.—There is, however, one method of applying water in cases of uterine hæmorrhage, which appears to be practical and scientific. It consists in the application of cloths saturated with water as hot as can be borne without pain, applied to the lower lumbar region. This plan was first recommended by an eminent English physician,* on the ground that *hot* applications to the spine cause contractions of the arteries supplied by the nerves of that region of the spinal cord. It also causes contractions of the uterus in the same manner as dry cupping applied to the lumbar region. The hot applications should be changed very often, and the loss of heat prevented by a covering of oiled silk.

I have tested this plan in several cases of uterine hæmorrhage, even before the placenta was expelled. It seemed to cause contractions of the uterus, which resulted in the arrest of the flooding, and in some cases in expulsion of the after-birth. In one case of advanced uterine cancer, where the floodings were frequent and copious, they were always quickly arrested by the application of napkins wrung out of hot water, and applied to the lumbar region.

It must be borne in mind, that all the above medicinal and mechanical means have only been resorted to because the os uteri was not *dilated* or *dilatable*; for as soon as this condition of the os is present, we are not justified in depending on the above-mentioned means.

The rule to be adopted in all cases of abortion, is: as soon as the os is dilatable or dilated, so as to admit the index finger, the blunt hook or intra-uterine placental forceps should be used for the extraction of the placenta, or such other portions of the embryo as may be in the uterus.

In the early part of my practice I adopted this rule, and have always found it the best, and that any deviation from it resulted in a protracted and troublesome case, unless nature expelled the offending substances.

* Braithwaite, 1863.

The *blunt hook* is the most generally useful instrument. It should be about 15 or 18 inches in length, with the hook at one extremity. An extempore hook may, in case of emergency, be made with a piece of strong wire.

BLUNT HOOK.

The following is the best method of using the hook. The woman lying on her back, with the knees drawn up, and thighs placed a proper distance apart (the patient may lie on the side of the bed, diagonally, or at right angles with the length of the bed, as the physician may choose); pass the index finger of the right hand (or left, if most convenient) up to the os uteri; ascertain if the finger can be passed into the uterine cavity; if so, pass the blunt hook along the palmar surface of the finger until it reaches the dilated os. The instrument is then to be carefully introduced into the interior of the uterus, unless the placenta is found protruding from the os, or engaged in the canal of the cervix, when it is easily removed. The hook once in the uterine cavity, we may feel for the placenta just as we would feel for a cork in a bottle, in the dark. Any one experienced in the use of the hook will readily detect when it comes in contact with the placenta. When this occurs, slowly and carefully draw it towards the cervix. When it arrives at that point, we should try to bring it in contact with the end of the index finger, which perhaps has not been withdrawn. With the point of the finger pressing against one side of the mass, the hook is readily forced into the other, and the whole extracted easily. Should the placenta prove too fragile, it will have to be extracted in pieces. A little patience and tact will enable the physician to get the whole away.

It is often difficult, and even quite impossible, to extract the placenta whole, or in fragments, **if it is very fragile.** The ordinary blunt **hook** will **not** hold in the tissue, but upon the slightest traction tears through; and this will occur again and again, making the removal of the mass a very tedious process.

In view of this difficulty, I have for several years used an instrument of my own invention, which has many advantages over the common hook. The bent extremity, instead of consisting of a single wire, is *looped,* as will be seen in the accompanying cut. The loop is about three-eighths of an inch across, and this allows the instrument to get a better hold upon the placenta, and traction will **not force it** through the fragile tissue. It is to be introduced **and used exactly in** the same manner **as the blunt hook.**

LOOPED HOOK.

The Forceps.—The intra-uterine placental forceps is a valuable instrument in some cases. Although not as generally manageable and efficient as the blunt or looped hook, there are many cases where it can be used easily. It differs from the ordinary placental forceps in having a longer curve, as will be seen by the accompanying figure. I do not consider the short forceps of any value except in cases where the placenta is protruding from the os, or lies in the vagina, and here the hook will answer every purpose.

LONG FORCEPS.

The method of introducing the long forceps is the same as advised for the blunt hook. The operator should be very careful or he will pinch the mucous membrane of the vagina or vulva, thereby causing the woman much pain. When the blades of the forceps are in the cavity of the uterus, the instrument should be carefully opened and closed, moving it before each closure, to a different portion of the interior. When the placenta has been seized, the fact will be known by the *handles of the forceps not coming in contact*. When this occurs, carefully withdraw the instrument, following the axis of the strait, and the whole or a portion of the placenta will be discovered by the finger in the vagina, between the blades of the forceps. If only a portion is extracted, renew the attempts until the whole is brought away.

If the os uteri is well dilated, and the placenta is felt engaging in the cervix, or presenting at the os, or even remains in the cavity of the uterus which lies low in the

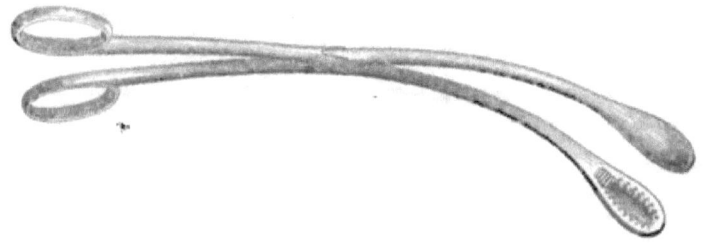

SHORT FORCEPS.

pelvis, then the short forceps may be used with advantage. The directions given for the use of the long, or intra-uterine forceps apply to this instrument.

The hook and forceps may be used at any period of pregnancy, and always as soon as the condition of the os will permit their introduction. At almost any period, their presence in the womb excites it to contractile efforts, so that there is little or no hæmorrhage at the time of their use. Should we fail to seize the placenta in a reasonable time, the tampon should be used, proper remedies administered, and the case left for twenty-four hours.

It often happens that while previous to the use of the hook, forceps or tampon, uterine pains were absent, they afterwards appear and become expulsive. If, therefore, after the tampon has been placed in the vagina, labor-like pains set in, and after increasing for a time in intensity, *suddenly cease*, we are warranted in believing the placenta has been expelled. We may then remove the tampon, and in the majority of cases, the placenta will be found in the vagina. We may, however, find only a large *clot*, or the fœtus, which in passing out of the uterus has caused the above symptoms, but after the expulsion of a clot or fœtus, the pains commence again after a few hours, or less time; whereas after the expulsion of the placenta, pains do not return, unless the patient is beyond the fifth month, when they are really *after-pains*.

It has been mentioned, that in cases of abortion before the sixth month, the danger from hæmorrhage is *before* the expulsion of the placenta and membranes; while after that date dangerous hæmorrhage usually occurs *after* the after-birth has been expelled. If this statement be fixed in the mind, it will be a valuable guide to the practitioner in the treatment of that accident.

Before the sixth month *medicinal* remedies cannot permanently check hæmorrhage previous to the discharge of the placenta. They may, for a time, exercise their specific effect over the bleeding vessels, but if the cause of irritation remains, the hæmorrhage will soon return. Remedies like *secale*, or *caulophyllum*, will cause the uterus to contract firmly upon the placenta, or force it into the canal of the cervix, but if they do not effect its expulsion, the flooding will return. There is no safety for the woman until the whole product of conception has passed out of the uterus.

After the sixth month severe hæmorrhage rarely occurs while the placenta remains in the womb, and it is the want of proper contraction in that organ which causes the flooding. In this case, medicines may be used to advantage, in connexion with the application of cold or hot water, as before mentioned, but the tampon should *never be used*. The remedies most generally useful in post-partum hæmorrhage after the sixth month, are, *Secale, caulophyllum, macrotin, erigeron, ipecac*, etc.

There are other measures which may be resorted to in dangerous and severe cases.

Dry cupping over the lower lumbar vertebræ, is said to cause firm and persistent uterine contractions.

Galvanism, when applied in a proper manner, has been successfully used in cases of alarming hæmorrhage. One pole should be applied to the os uteri, or pubes— the other to the sacral or lower lumbar region, and an interrupted current passed through the uterus. Firm contraction is said to immediately ensue.

The same means adopted for the arrest of hæmorrhage during the last three months of pregnancy, are the most useful ones to facilitate the expulsion of the contents of the uterus. The same contractions which expel the placenta will prove efficient for the prevention of flooding, if they continue for a proper length of time.

As regards the management of the after-birth (and by after-birth is here meant all decidual or placental substances left after the expulsion of the embryo or fœtus), it will be considered in a separate chapter.

SECTION VIII.

CONDUCT OF THE PHYSICIAN.

The medical man who is called upon to **attend a case** of Abortion, occupies a peculiar and anomalous position. There are three classes of cases with which he has to deal, and each requires a different method of action **on** his part. The *first*, are married women of respectability, in **whom the a**bortion has been brought **on by** accident, or some of the diseases mentioned among the causes enumerated. In these instances there is usually no hesitation on the part of the patient in admitting to the physician the nature of the illness, and she may not have any objection **to its** publicity among her friends or neighbors. In some cases, however, women of sensitive minds are averse to having a knowledge of the character of the sickness made known even to their nearest friends. In these cases, the evident duty of the physician is, if he has timely notice, to use all the means at his disposal to save the life of the child and arrest the progress of abortion, if he can do so with perfect safety to the mother; or, if called too late, he must exercise his highest skill in conducting the abortion to a safe conclusion, and afterwards proceed to effect a cure of the disease (if any existed) which acted as a cause of the accident.

The *second* class comprises married women who entertain wrong ideas concerning their duty, and imagine that lack of means, ill health, or other adverse circumstances are sufficient reasons why they should not bear children. This, together with a wish to shirk the trials and responsibilities of maternity, as well as a desire to lead easy lives, or lives of fashion and luxury, leads them to take measures to prevent the fruit of conception from reaching maturity. To accomplish this they resort to quacks, charlatans, or unprincipled women, who are accustomed to induce abortion; or use the various nostrums so shamelessly advertised in many periodicals, or the instruments which are in the hands of so very many of the women of this country. The abortion produced by any of the above means, they attempt to endure the suffering, and risk the danger, rather than call a physician and thereby have their criminal conduct known to the medical attendant, or their own family. Such patients, when they do call a physician, attempt to hide from him the real nature of the malady, and hope that he will treat the case successfully without a knowledge of the actual nature of the case. They will call it a dysmenorrhœa, an injury from a fall, etc., or a simple menorrhagia, and it requires considerable tact to ascertain from them the facts.

In these cases, the physician should come to no hasty conclusion, married women rarely seek aid for an attack of dysmenorrhœa or menorrhagia, and is is a good rule to suspect the occurrence of an abortion when called to attend such alleged cases. If the woman denies that there is any hæmorrhage, as they will often do, or declares the attack to be colic or dysentery, then the trouble of detecting the real disease is much more difficult. But the facts being ascertained, (see "Examination of the Patient,") the causes of the abortion, as well as the symptoms present should all be taken into account, and the case treated as above advised.

The *third* class are those unfortunate women, married and unmarried, the victims of the seducer, or of their own perverted passions, who have stepped aside from the paths of virtue, and with the frailty which pertains to humanity, transgressed the laws of God and man. In such cases the honorable physician is placed in a very delicate and trying position. The patient has perhaps applied to him to rid her of the consequences of her sinful conduct, but upon his refusal to do a criminal act she has resorted to some of the unlawful means mentioned above, and brought about the impending or actual abortion. In such cases it is difficult to decide what should be the conduct of a physician in one respect: namely, should he attempt to arrest the abortion, and thereby bring the mother to inevitable shame and save the life of an illegitimate child, whose unnatural parentage will be a disgrace all through life? Probably the best and most consistent conduct for the physician to adopt, should be to simply let outraged nature take her own course, and only interfere by warding off those symptoms which threaten the life of the mother.

In relation to the conduct of the physician during and after the abortion I have a few suggestions to make. There is a great lack of the proper honor and delicacy, regarding a mention of such accidents, both with physicians and others. It is common for nurses, friends and physicians to say, that "Mrs. A. or B. has '*slipped up*,'" or to allude to the accident in some such vulgar manner. The unwarranted impertinence of neighbors, who insist upon knowing the nature of every woman's illness, is another reprehensible habit. It forces the physician to equivocate or divulge his patient's secret.

The rule which should be invariably adopted by every medical man, is this: *Whether in or out of his patient's room, during her illness or after, no matter what her condition in life, character or standing in society, in short, under any circumstances whatever, he should avoid all mention of the occurrence of an abortion in his patient.* If people are so devoid of good breeding as to ask impertinent questions, he is justified in equivocating to the extent of blinding the questioner.

SECTION IX.

EXAMINATION OF THE PATIENT.

To physicians who have had much experience in the treatment of abortion, I need not dilate on the importance of ascertaining as soon as possible that an abortion is impending or in progress, nor need I mention the almost insuperable difficulty which exists in some cases, of satisfying ourselves during our first visits, whether the above accident is what we are called to treat.

Patients belonging to the *first* class previously mentioned, rarely cause us any trouble, but will frankly say to the medical attendant, "Doctor, I am threatened with an Abortion and would like to have you arrest it if possible," or "I am having a miscarriage." This class will often intelligently acquaint us with the history of the case, the present and previous symptoms, and submit sensibly to all necessary examinations for the purpose of ascertaining the real condition. As there is always more or less nervous agitation attending such cases, manifesting itself especially when the physician is present, a few directions may not be amiss to the student or young practitioner.

As it is customary for the messenger, if it is some relative, to acquaint the physician with the nature of the illness, it is to be supposed he will learn something of the nature of the case, either with or without direct questioning, in which case he will feel somewhat prepared to meet his patient. Upon entering the room, after some brief salutations, he should quietly take a seat by the bedside of the woman, and place his fingers upon her wrist. Meanwhile the state of the pulse and the appearance of the patient may be noted. If the pulse is weak or thready and the face and lips blanched, it is presumed that severe hæmorrhage has occurred; if, on the contrary, the pulse is full, hard or normal, and the face red or natural, there is irritative fever, general reaction, or only the premonitions of abortion, or actual abortion in the early months. We must not, however, judge too hastily, as the mental excitement of the patient may cause a suffused face and quick pulse.

Generally the first appropriate question to the patient or her nurse, is: " When did this illness begin ?" next "What have been the symptoms until now ?" There are instances, however, where the evident exsanguification should impel us to ask at once, "How long has she been flooding ?" In some cases even, it will be found absolutely necessary, without asking a single question, to state to the patient or nurse what must be done, call for material for a tampon and use it immediately, and then ply the patient with active stimulants.

So soon as the immediate danger is over, or an abortion is ascertained to be progressing, the physician should inquire if the fœtus and placenta have been expelled. This is a point upon which the closest investigation is necessary. We are often informed, that " every thing has come away," when in fact the patient or her nurse has seen only the fœtus and clots of blood, or perhaps nothing but a mass or masses of coagula. Not one patient or nurse in a thousand ever examines sufficiently the expelled substances. The fœtus and after birth, at an early stage, may be passed, enveloped in large coagula, and altogether escape superficial scrutiny, or neither may be present, yet are supposed to be from the general appearance of the mass.

It often becomes necessary to have the substances extruded, produced, that we may examine carefully their character. Each clot should be broken and its interior examined for embryo or placenta. If we cannot satisfy ourselves on this point, we must rest in uncertainty until an examination can be made. If one fact, however, is borne in mind, it will greatly aid us in forming our opinion, namely, *before the sixth month hæmorrhage rarely, if ever, occurs, after the placenta is expelled.* The expulsion of the fœtus *alone*, only causes a subsidence of the pains for a time; it does not have any influence over the arrest of hæmorrhage.

We have supposed cases where the abortion was admitted, or was evident at first.

The *second and third* classes are more difficult to treat,

because of the obstacles thrown in the way of a proper knowledge of the case.

A physician is called to see a woman. He has no intimation of the nature of her illness. She is said to be in "great pain" or "very sick." She says to him, "I want relief from the pain across me," or "I have a colic." She will state it is from a cold, and assert that it was from a suppression, or intimate that she has been regular, and the attack is only dysmenorrhœal. A woman who has determined to hide her shame will use every subterfuge that cunning and equivocation can devise. She will deny the presence of flooding, even when her exsanguined appearance plainly indicates the fact. The character of the pain will be an indication to the watchful physician. But an ordinary colic is often intermitting, as also are the pains of dysmenorrhœa or dysentery.

The only way we can, in some cases, arrive at a correct diagnosis, is to *seem* to assent to the assertions of the patient, while we are questioning in relation to symptoms which, while they seem unimportant to her, are valuable indications to us. When we are satisfied than an abortion is in progress or impending, we should not hesitate to acquaint the patient with our convictions, in a courteous but decided manner. Instances are not rare, where careless or ignorant physicians have attended cases of abortion during the first half of gestation, under the idea that they were cases of enteralgia, enteritis, dysmenorrhœa, or even dysentery; the woman succeeding perfectly in hiding the evidences of the miscarriage. Every physician should be on the alert and not allow himself to be caught in such a mortifying predicament. Finally, in an examination of a patient, it should be remembered, that no matter what her social condition, her character, or the character of those around her, the sick-room is no place for unseemly or rude jests. No coarse allusions or indelicate remarks should be indulged in or *allowed* by the physician. No one having any regard for manly or professional dignity will countenance such conduct. All manual examinations should be conducted with delicacy and gentleness, without any unnecessary exposure of the patient. If instruments are used, all unnecessary flourish and exhibition should be avoided, for these are the tricks of quacks and ignorant pretenders.

SECTION X.

MANAGEMENT OF LABOR.

"Abortion," says Professor Hodge, "is a true labor. The two characteristic peculiarities are—first, that there is no placenta, but the whole membrana decidua is in close and vascular connection with the internal surface of the uterus."* .Dr. Hodge restricts the term abortion to an expulsion of the ovum before the *fourth* month, after which period the placenta is formed, and the decidual membranes are separated from the uterine surfaces.

The practitioner, when called to attend an abortion, should go prepared to meet any emergency which may arise. If he neglects this, and the woman dies, or suffers from dangerous symptoms which might have been remedied by the proper medicines and instruments, he is culpable in the highest degree for his criminal carelessness or ignorance.

* System of Obstetrics, page 464.

He should take with him, besides his ordinary pocket-case of remedies, some crude *ergot* freshly prepared, or a reliable vinous tincture; or the first decimal trituration of *caulophyllin;* some *oil of erigeron*, $\frac{1}{5}$; a blunt hook, or abortion forceps; and a piece of charpie, or a sponge, to use as a tampon. If the physician has a well-regulated office, all these can be collected in a moment. Their possession may be of invaluable service to him, and enable him to save the life of his patient, and his own reputation, both of which might be lost while a messenger was riding miles away to procure the articles desired.

The treatment of the premonitory symptoms has already been laid down; also the treatment of the symptoms, such as hæmorrhage, etc., which may occur during the progress of an abortion.

The *management* proper of the different stages, will now be considered.

If, upon examination, the os is found closed, no hæmorrhage, and no change taken place in the body of the uterus, except some descent of that organ in the pelvis, and some unusual turgescence and fullness of the pelvic viscera, such as simple organic irritation would cause; the physician can do no more than to prescribe the appropriate remedy for the condition, and give the proper directions relating to diet and posture, and leave the patient for the time. (See Treatment, page 218). If, however, nervous irritation or contraction has ensued, and the neck of the uterus is found developed at the upper part, it will indicate a partial descent of the ovum into the canal of the cervix. If this occurs, the conical form of the lower portion of the uterus will be found considerably altered—it will have become more spherical. Dr. Lee states that, in all such cases, expulsion will surely follow. This termination will be still more probable if, in addition, the os uteri is found partially dilated, and especially if a portion of the ovum be found protruding. There are cases, however, especially in multiparous women, where the internal os remains contracted, while the external is soft and somewhat patulous. In these instances, if alternate pains be not severe, there is ground for hope that the ovum will be retained.

If, with the above condition, there is still no hæmorrhage, the patient may be left as before, under the use of the remedies, etc., appropriate to the symptoms and state. The *remedies* indicated are—*gelseminum, cimicifuga, caulophyllum, gossipium* and *secale*, from the 3rd dilution upwards. It is in this condition that full doses of *opium* have been found most useful by the old school, and also even when hæmorrhage is present. (See case, page 223).

The best *position* for the patient to assume, is on the back, with the hips elevated, unless in case of retroversion, when the woman should lie on the face. Again, if it be positively ascertained that the ovum is blighted, that the liquor amnii is evacuated,* that the hæmorrhage is profuse, and regular alternating pains are established, and especially if the measures for prevention have been inefficient, the practitioner should favor, not retard, the expulsive process; and, at the same time, render it as easy and as safe as possible.

In this place I must advert to my division of uterogestation into *three* stages, for the management of this expulsive stage of abortion, must be varied with each stage of pregnancy.

---

* This occurrence is rare at the early months, as, in a great number of cases, the ovum is discharged entire; and in all instances the quantity of the amniotic fluid is so small that, when discharged, it escapes notice.

In the *first* stage, it is rare that we find much dilatation of the os uteri, or can find the embryo presenting. Should we detect the embryo engaged in the canal of the cervix, or even in the vagina, it is not best to remove it, for its presence in either canal may be useful in exciting uterine contractions; while its removal can do no good, but rather retard the progress of labor. Its presence in the cervical canal may also be useful in preventing hæmorrhage, acting as a plug. We should not leave the woman in this condition, without placing a tampon *in situ*, for fear that hæmorrhage occurs in our absence. The presence of the tampon will be of service, too, in hastening the termination of the expulsive process.

The delivery of the after-birth, or secundines, which, in this stage of pregnancy, consist only of the decidual membranes, is the most important duty of the physician. It is difficult to accomplish this with the ordinary wire crochet, or blunt hook, owing to the delicacy of the membranes, and their general adherence to the uterine walls. Neither can we rely upon those uterine-motor medicines (*caulophyllin, ergot*, etc.) for, in this stage, the contractile power of the uterus is small, owing to the undeveloped condition of its muscular tissue.

We must either allow them—if not thrown off by the efforts of nature—to decay, and pass away in a fluid state, or extract them by instrumental means other than the hook. The intra-uterine placental forceps, to which I have alluded on another page, may be used; or the " abortion forceps," originally recommended by Levret, or Hodge's forceps, " the blades of which," he says, " are so arranged that they may present the form of a lever, allowing their easy introduction through the cervix into the uterus, exterior to the ovum." (It is only in the third month, or after, however, that such an an instrument can be used.) But, probably, the instrument best adapted to remove the contents of the uterus, if the os and cervix are dilated so as to admit the finger, is the "abortion vectis," which I have invented. It is, it is true, only a modification of Hodge's "lever and crotchet," but it is specially adapted for the removal of the embryo in this *first* stage. The lever and crotchet combined, is better adapted for use in the *second* stage.

ABORTION VECTIS.

The abortion vectis can be insinuated into the uterus, betwen the embryo and the uterine parietes, and, with careful oscillating movements, we may separate the ovum from its adhesions, and generally withdraw it entire. So soon as this is accomplished, hæmorrhage and pain will cease, and the abortion is at an end. If, however, even a small portion of the membrane is left in the womb, it will, in certain constitutions, act as a local irritant, keeping up pain and hæmorrhage, until it has disintegrated and passed off; or, its absorption may cause irritative fever, which must be treated with *sulphite of soda, baptisia, arsenicum*, or *chlorate of potash*, internally, together with injections of *chlorate* or *per-manganate of potash*, or *baptisia*, into the interior of the uterus, to wash away the offending substance, and act as a local disinfectant. The instrument best adapted for this purpose is a hard rubber syringe, holding two ounces, and provided with an extra long and slender pipe.

If the abortion occurs during the *second* stage of pregnancy, the management of the labor differs somewhat from that of the first stage. The delivery of the embryo-fœtus should be left to nature, as recommended in the first stage, and for the same reasons, the delivery of the placenta and membranes may have to be assisted in a somewhat different manner. After the third month the placenta is perfect, and is attached to the uterus, while at the same time the decidua ceases to be in such contact with the uterine wall as in the previous stage, the circulation between the mother and child being carried on solely through the placenta.

In this stage, abortion may occur with complete detachment of the whole placenta at once, in which case the embryonic mass drops into the lower segment of the uterus, and may be expelled entire, as in the first stage. Oftener, however, the membranes are ruptured, and the fœtus escapes while the placenta retains its connections. In other cases we may have abnormal adherence of the placenta, in which instance the contractions of the uterus, though sometimes quite powerful, are insufficient to detach it. Here the "abortion vectis" will form a valuable instrument, if the os is open sufficiently to admit of its introduction. After it has been detached, however, the blunt or looped hook or forceps will form the best instrument for its extraction.

If the os is not sufficiently dilated we may use the tampon, and give *gelseminum* or *belladonna* until dilation occurs. We may then have to administer *caulophyllin*, or *ergot*, the former in doses of two to four grains of the $\frac{1}{18}$th, the latter in the officinal dose given from time immemorial—namely, one drachm to a teacup of hot water, taken at four draughts fifteen minutes apart. Those who think the 3rd or 30th attenuation will answer as well are allowed their opinion.

An adherent placenta, remaining in spite of the efforts of nature or art, is supposed to be the origin of molar or hydatid masses. But it quite often happens that the whole placenta undergoes putrefactive disintegration, or the portions left after instrumental interference, decay; in which case we may have fœtid discharges and irritative fever, previously alluded to and treatment given.

The management of labor after the sixth month (*third* stage of pregnancy), is essentially the same as recommended for labor at full time (see standard works on Obstetrics). I will only add a few observations to refresh the memory of the practitioner, namely:

The expulsion of the ovum may here be aided by the internal use of uterine motor excitants; by instruments; and by remedies Homœopathic to non-relaxation of the os and soft parts.

The delivery of the placenta may be aided by traction with the cord (which cannot be done in the earlier months); by *caulophyllin, ergot*, etc.; and by the instruments heretofore named.

The hæmorrhage must be arrested with *ergot*, cold water, galvanism, dry cupping, *erigeron*, the tight bandage, etc., as recommended in cases of natural labor.

Here the *tampon should never be used*, nor those remedies recommended for hæmorrhage during the first stage. They are not often adapted to the pathological state upon which post-partum flooding depends, and will generally prove insufficient.

SECTION XI.

SEQUELÆ OF ABORTION.

The *post-partum* treatment of abortion, although not of the same importance as the *remedial*, is a subject which should be fully considered. I believe it has not received that attention from physicians which the subject demands.

The *post-partum* treatment may be properly divided into

    (*a*) *Postural.*
    (*b*) *Dietetic.*
    (*c*) *Medicinal.*
    (*e*) *Mechanical.*

The *postural* treatment is altogether too much neglected by a great majority of physicians.

It is too often the case that a patient who has had an abortion during the first, or even second stage of pregnancy, is allowed by the attendant to rise from her bed the next day, or a few days thereafter, and attend to her household duties or recreations. I have known women engage in severe work or exercise, or attend parties or balls, but a few days after an abortion in the second month. The physician has a responsibility in this matter. It is his duty to state to the patient, her nurse, or certain members of the family, the importance of the recumbent posture for a certain length of time after the delivery. This period varies somewhat with the period of pregnancy in which the abortion has occurred.

It depends, too, in a great measure, upon the severity of the symptoms during the progress of the abortion, and upon the constitution and morbid conditions of the patient.

Abortion in the first stage of pregnancy is rarely attended with great loss of blood, the pain is not very severe, and consequently, the strength of the woman is not much exhausted. But an engorged condition of the uterus is generally present, and if the woman is too soon and too often on her feet, or sits up too much, the womb is inclined to sink into the pelvis, and a prolapsus or retroversion ensues.

If the abortion has been short in its duration, and the symptoms few and not severe, confinement to the bed or couch need not continue more than four or five days. If, however, getting up and standing or walking for an hour or two causes any local uneasiness in the back or abdomen, the recumbent posture should be maintained a portion of the time for several days longer.

If, on the contrary, the symptoms have been severe; there has been much hæmorrhage; a slow expulsion of the embryo, or its retention in utero; or if it is found on examination that the uterus or its cervix is congested, or the os lacerated or abraded, explicit directions should be given not to allow the patient to sit up or stand on her feet more than a minute or two at a time until eight or ten days have expired. The woman who leaves her bed under these circumstances before that time, does so at the imminent hazard of her future health, and perhaps her life.

During the second period, the postural treatment of patients who have aborted is of still greater importance. As it approaches the sixth month, confinement becomes more and more hazardous, and its sequelæ more like those of labor at term.

Hæmorrhage, when it occurs, is more apt to be profuse and exhausting, the labor more tedious and dangerous, and the after treatment more important. It is believed by practical and observing physicians that cancer of the uterus, ulceration of the os and cervix, chronic metritis, prolapsus and retroversion, frequently occur from a too early assumption of the erect position after abortion in the middle period of gestation. In nearly every case, should the symptoms appear ever so slight and unimportant, the woman should keep her bed for a week after the labor has been completed. There are certain cases, however, where the placenta is retained; but the patient improves in strength and health notwithstanding. The placenta may not be expelled for weeks after the woman has left her bed, no decay or putrefaction having taken place. In these cases, we cannot, of course, confine the patient to her bed until its expulsion occurs, nor would it be proper to do so.

If severe hæmorrhage, congestion, or inflammation, or irritative fever has occurred, the patient should on no account get up until the twelfth day, even if she feels perfectly able to do so. In these instances, however, the woman has not the power of rising, even if she desires to, such is the local pain, general weakness, vertigo, etc., which is caused by the effort.

If the accident has occurred during the last three months of pregnancy, nearly the same rules, last mentioned, should govern us in our postural treatment. The same directions should be given as would be proper after delivery at the ninth month, as the pathological conditions which ensue are nearly the same as at the latter period.

Dietetic.—The diet of the patient who has suffered an abortion is not an unimportant item. Its character will depend on the condition of the system. If fever is present, due to any local inflammation, the aliment should be very light, easily assimilable, and bland. Toast water, rice gruel, flour or farina porridge, light toast or crackers, together with cooling beverages, will be best relished by the patient, and most appropriate. If the fever is irritative, from local irritation or the absorption of morbid matters into the circulation, the food allowed should be of a more nutritious character. Milk, rice, toast, animal broths, beef-tea, wine and whey, cocoa, etc., may be allowed. The physician should be careful to form a correct diagnosis in these cases, and not mistake the *irritable* pulse of this condition for the quick, hard pulse of fever from inflammation.

If the patient has lost much blood, or been greatly reduced by the length of the illness, or prostrated by the irritative fever which has attended it, the diet will have to be not only exceedingly nutritious, but sufficiently stimulating. We shall be under the necessity of ordering milk-punch, egg-nogg, beef-tea, mutton-broth, soups, all such meats as the patient can relish, and allow her to take them in such quantities as her condition requires. It is very necessary in these cases to give the nutriment at short intervals. If food is given every two or three hours, the quantity taken will be less, and the stomach is not so liable to be overloaded. Dr. Inman states that he has seen cases where it was necessary to administer to the patient concentrated nutriment every hour, in order to prevent dangerous prostration. It is hardly necessary to caution against allowing indigestible articles of diet, among which may be mentioned raw apples or other uncooked fruits, pickles, salt meats or fish, pastry, confectionery, etc. In this category might be placed ice-cream, were it not that this article is often allowable, especially in cases of great prostration or irritative fever, where there is present great thirst, dryness of the mouth, and a desire for cooling drinks. It may be given in small quantities, a spoonful at a time, and allowed to be slowly melted in the mouth. If taken in too large quantities, or too hastily, it may give rise to colic and even spasms.

The *medicinal* treatment of the sequelæ of abortion will depend upon the nature of the pathological condition, and the symptoms present.

These sequelæ have, the most of them, been described, and their treatment given, in that part of this work devoted to the causes and their treatment. There are other maladies, which are not mentioned; but which properly belong to the consequences of abortion. They are, pelvic cellulitis; hypertrophy of the uterus; fistulæ, (vesical, uterine, or between the organs alluded to); adhesions of the os and vagina, rendering liable subsequent rupture of the womb during labor or from retained menses, or, in the latter case, discharge of the secretion through the Fallopian tubes and consequent peritonitis; and, finally, inflammation of the mammæ, puerperal metritis, phlebitis, dropsy, and paralysis of the lower extremities. Others might be named; but it would swell the list to an extent incompatible with the limits of this treatise.

The medicinal treatment of these sequelæ will be briefly alluded to.

Pelvic cellulitis will require those remedies which favor resolution or healthy suppuration, namely: *Mercurius, belladonna, hepar sulph.*, and *phytolacca*. In some cases the abscess will have to be punctured to give vent to the pus, which, if confined, would burrow, and create troublesome complications.

Hypertrophy of the Uterus.—I have observed in many cases of abortion before the sixth month, a large increase in the size of the abdomen in the hypogastric region. It was not ascites, for external examination failed to detect the fluctuation found in dropsy, and moreover the urinary secretion was normal in quantity. An examination of the uterus showed it to be much enlarged—often much larger than before the abortion. I believe this to be a serous infiltration into the uterine parieties, superadded to that condition which Professor Simpson has described as "Sub-involution of the Uterus."

I have treated such cases successfully with *sepia, arsenicum, kali bromatum* and *kali hydriodicum*.

Fistulæ, adhesions, etc., will require the attention of the skillful surgeon; but they may be prevented by proper attention on the part of the physician, who, if he has the least fear of such consequences, should make occasional examinations, and by placing a dilator in the os, or pledgets of lint in the vagina, prevent such evils.

Inflammation of the Mammæ. When the abortion has occurred after the third month, it is not unusual for milk to be secreted. When this happens it will be necessary to arrest the secretion as soon as it can be safely done. This end is best attained by prescribing, for internal remedies, *kali hydriodatum, phosphorus* and *magnesia sulphurica*; and externally embrocations of *aconite-water* (℥j to ℥j), *phytolacca, belladonna*, or *iodide of potash*, in ointment or solution. If, from neglect, or in spite of these means, inflammation and abscess supervene, adopt the treatment herein-before recommended.

Puerperal metritis will require *veratrum viride, belladonna, stramonium, cimicifuga*, and external applications of *aconite*.

Puerperal peritonitis, the same remedies. (This disease is not of such frequent occurrence as is believed. It is often confounded with *myalgia*, which requires *caulophyllum, nux vomica, cimicifuga, helonias, china bryonia*, and *colocynth*, with external applications of a lotion of *gelseminum* or *arnica*).

Phlebitis calls for *arnica, belladonna*, and *hamamelis*, — the latter externally applied if possible.

Dropsy after abortion is removed by *apocynum cannabinum, eupatorium purpureum, apis mel., cantharis, senecio aureus*, and *arsenicum*.

Paralysis of the lower extremities will yield to *caulophyllum, hedeoma, nux vomica, ignatia, citrate of iron and strychnia, secale*, and *æsculus glabra*.

Mental aberrations are not uncommon after abortions, and may proceed from reflex influences, or from grief or remorse. These will require careful symptomatic treatment. One of the best remedies for the *mental depression* following abortion is the *cimicifuga*, which has been successfully used by Professor Simpson, and lately by myself.

One of the most common of the sequelæ of abortion is *premature* and *profuse* menses. They sometimes occur every three weeks; sometimes every two weeks. Although *sabina, ipecac, sulphur*, and *platinum* are frequently indicated and successful, I have been most successful, in fact in nearly every case, with *senecin* and *calcarea carb.* In these, as in some other instances in our practice, neither seems to be as useful alone, as when used in alternation.

I usually prescribe the *senecin* in the second decimal trituration, a dose every afternoon and evening, and a single dose of *calcarea*, third trituration, every morning.

It is best to give these remedies, beginning just after the last menstrual period, and continuing them until the next regular period, unless the menses are premature, in which case they may be superseded by *ipecac* or *sabina* during the flow.

It is rare that the second period will be premature after beginning the use of the above remedies.

Chronic menorrhagia of a very obstinate character sometimes follows an abortion. It may arise from atony of the uterus, from irritability, or from a condition of the lining membrane, described by some writers as consisting of an enlarged and hypertrophied state of the villous coat, — a kind of fungous growth of its papillæ, from which blood oozes upon the least irritation.

For atonic conditions, we should use *helonias, trillium, secale, aletris, terebinth, sulphuric acid,* and *sabina,* in the lower dilutions.

For irritation, *sabina, caulophyllum, platina,* etc. The last cause of *menorrhagia* can only be cured by absolute rest and a long course of treatment, with carefully selected remedies, among which may be mentioned *erigeron, secale, trillium, calc., senecio, china, ferrum, pulsatilla,* etc., unless we resort to local applications to the diseased surface, applied with the intra-uterine syringe.

Several cases have occurred in my practice, which persisted for many months, notwithstanding the use of all the most approved medicines recommended for *menorrhagia*, until topical means were resorted to. One case I cured with injections of *tinctura ferr. muriatis*, 10 drops to ℥j of water, half of which was used once a day. After the third day, the hæmorrhage ceased, and did not return. In another case the *pernitrate of iron*, was used successfully.

The *mechanical* treatment includes those appliances which will serve to prevent the occurrence of uterine displacements, or the closure of the vagina or os uteri by adhesions.

A *prolapsus* may be prevented or cured by the same means, together with cool injections containing *hamamelis, nux vomica*, or *helonias*, the use of cool hip baths, and the hypogastric bandage.

For the prevention of adhesions, the vagina and uterus should be examined, in cases where excoriation or laceration is supposed to have occurred, to ascertain if the vaginal and cervical canals are normally open. The finger will do to explore the former, the sound the latter. If a closure is to be found, pledgets of lint, or pieces of linen spread with *calendula cerate* on both sides, should be introduced and kept *in situ* till the danger has passed by.

For *retroversion* I do not hesitate to use the closed lever pessary or the ring, as seems best indicated, and have thus prevented in many cases the uterus from assuming a retroverted condition, which it had attained previous to the abortion.

In my pamphlet on Retroversion and Retroflexion of the Uterus (page 2), I alluded to the fact that that form of displacement was a frequent cause, accompaniment, and consequence of abortion. I advert to it again because the subject is too important to be lost sight of by the practical physician.

As a *cause* of abortion, it has been fully treated of in the foregoing pages. As an *accompaniment*, I did not mention it to the extent I intended to do. A flexed or retroverted uterus is one of the chiefest obstacles to the natural expulsion or extraction of the placenta or any of the products of an arrested conception. Coagula of blood even are sometimes retained a long time in the cavity of the uterus from this cause alone.

When the fœtus dies, or the membrane becomes detached from the womb, that organ sinks in the pelvis, and if any tendency to retroversion exists, that accident is pretty sure to occur. In this case the expelling power of the uterus is altogether destroyed, and it strives with unavailing efforts to throw off its contents. A single glance at the anatomical peculiarities of the uterus will show us how futile these efforts must be. A hollow, pear-shaped muscle is flexed at its neck, or almost doubled upon itself. In this condition the severest spasmodic action of its muscular tissue becomes utterly useless. I have known the most intense spasmodic pains to continue day after day, weary the patient out, and no good result until the uterus was replaced and kept in a normal position until the offending placenta was expelled. A case which occurred in my practice not long since, so aptly illustrates this subject, that I cannot forbear giving it. A delicate, nervous woman, who had had several miscarriages, attended with retroversion, for which she had been treated by several physicians of eminence in our school came under my care. On one occasion, she had lain several weeks with a retained placenta, from retroversion, suffering in the meantime from metritis, spasmodic pains, etc., until the uterus was held in its proper position by a ring pessary, when the placenta came away in a horribly putrid state of semi-liquid putrescency.

In the instance in which she came under my care, the fœtus (of three months) was expelled; the pains ceased for nearly a day, but returned severely after she had imprudently made considerable effort at stool. After the pains had lasted several hours, I was called, and found them very agonizing and spasmodic. On *examination*, the uterus was found almost completely retroverted. Not having with me a *sound*, I left some mother tincture of *caulophyllum*, ten drops to be taken every half hour until I returned. Not getting relief in two hours, she took, on her own responsibility, a *teaspoonful* of the tincture at a single dose. The pain ceased altogether in fifteen minutes! This action of *caulophyllum* should not surprise us, for it is *secondarily* Homœopathic to spasm, in which case we know it often requires massive doses to prove curative. On getting on her feet again, however, the pain returned with its original intensity. On my arrival, the uterine sound was with much difficulty introduced, and with considerable effort the uterus was lifted into normal position. Knowing that it would become retroverted immediately if the sound were withdrawn, I introduced a *curved* elastic pessary while the sound was still in the uterus. This can be done by any one familiar with the former instrument. The sound was then withdrawn, and the uterus elevated as much as possible. *Ergot* was given in the usual manner, and the placenta was expelled in less than two hours. All pain ceased, and the pessary was allowed to remain for eight days, when the patient was discharged.

There have been several cases, which I was obliged

to treat in the same manner, and in view of the uniform success which has attended the method, I do not hesitate to advise the insertion of the elastic *curved* or *ring* pessary, in every instance of retained placenta from retroversion.

It will sometimes happen that this form of displacement will not occur until after the abortion has fully terminated with the complete discharge of the fœtus, placenta, and membranes. The patient gets up too soon, and the effort of standing, or straining at stool, or lifting, causes the heavy fundus-uteri to fall backward. The attending physician should warn his patient of the nature and probable occurrence of the accident, and the means to avoid it. She should be instructed to report if certain symptoms obtain, that immediate measures for relief may be taken. As soon as the displacement is known to have occurred, the physician should enjoin the recumbent position, replace the organ by means of the processes hereinbefore described, and place the patient upon the use of *iodide of iron, podophyllum, nux vomica, helonias*, or *belladonna*, until the danger of a recurrence of the accident is passed.

PART V.

OBSTETRIC ABORTION.

i

SECTION I.

OBSTETRIC ABORTION.

It would be a manifest omission, and detract largely from the value of this work, were I to leave out of it a full consideration of the subject of obstetric abortion, namely: the conditions under which it is proper and necessary that labor should be brought on before the ninth month.

I shall proceed to treat of this subject in three divisions, which will correspond with the three divisions of utero-gestation which I have heretofore described.

First. The various methods in use to cause expulsion of the contents of the gravid uterus, after the sixth month, and before the time for the normal termination of pregnancy. This is designated as *premature labor.*

Second. The methods adopted to bring about the same results during the middle period of gestation, or from the end of the third to the end of the sixth month. This is termed *fœtal abortion.*

Third. The methods which may be used during the first three months after conception. This is *embryonic abortion.*

The abormal conditions which make it necessary to resort to these operations will be mentioned.

(1) *Premature Labor.*

The subject of *premature labor* cannot, in a work of this scope, be more than briefly mentioned. Every physician is supposed to have in his library the works of the most eminent obstetricians, in which will be found all the information required on this point.

Premature labor is an operation by which the lives of both mother and child may be saved. It is only admissible to resort to it in cases where the child cannot be born alive unless by the dreadful operations of the Cesarian section or symphyseotomy, both of which endanger the life of the mother in the highest degree; also in cases where delivery is only possible by embryulcia, or the extraction of the fœtus by fatal mutilation.

Premature labor is restricted to that period subsequent to the "viability" of the fœtus. This period is technically included within the last three months of pregnancy. There is but little hope, however, that a child, born in the sixth month of gestation, will survive, although cases are on record where children have lived at six months. At the seventh month children often live, and quite generally in the eighth.

The necessity for the induction of premature labor arises mainly from a *disproportion between the passages of the pelvis and the size of the head of the child at the full period of utero-gestation*. At this time, it may be presumed that the bi-parietal diameter of the fœtus measures at least three inches and four lines, and therefore a living child can hardly be born, even by aid of forceps, unless there be three and a quarter inches in the short diameter of the pelvis. Embryotomy, or some other operation, would be the alternative.

The same declaration applies to other sources of obstruction than simple deformity, as in cases of exostosis, of ovarian and other tumors in the pelvis, of indurated or contracted vagina, and in those also of permanent induration, thickening or scirrhous of the os and cervix uteri, or where there are fibrous or steatomous tumors in the lower segment of the uterus, so located or so large that the delivery at term would be impracticable. Also, it may be applicable in cases where successive pregnancies have demonstrated that the child is born dead in consequence of the size of the head, and its very complete state of ossification.

To estimate as nearly as possible at what period labor should be induced, the practitioner should examine carefully,

First, As far as practicable, the degree of deformity or obstruction which exists; and,

Secondly, The usual length of the bi-parietal diameter during the last three months of utero-gestation.

The modes of measuring the degree of deformity of the pelvis will be found detailed in the standard works on obstetrics.

As regards the diameters of the child's head, the following tables, prepared by M. Figueira and Ritgen, will afford a proximate idea of the length of the diameters at the different periods of gestation; although great allowances must be made for the relative development of different children, the uncertainty of the exact periods of conception, and, of course, of the stage of pregnancy, and for the flexibility or compressibility of the fœtal head, so that a cranium apparently too large may or can be often safely delivered.

Measurement of the bi-parietal diameter according to Figueira:

	In.	Lines.
At 7th month,	2	9
" 7½ "	3	–
" 8th "	3	1
" 8½ "	3	2
" 9th "	3	4

According to Ritgen, labor may be induced at the

								In.	L.
29th week, or	6½	month,	when the antero-posterior diam. is					2	7
20th "	6¾	"	"	"	"	"	"	2	8
31st "	7th	"	"	"	"	"	"	2	9
35th "	8th	"	"	"	"	"	"	2	10
36th "	8½	"	"	"	"	"	"	2	11
37th "	8¾	"	"	"	"	"	"	3	—

The best writers on obstetrics do not hesitate to assert, that, if the antero-posterior diameter of the brim be three inches and six lines, it will be justifiable to induce labor prior to the full period — that it will be better for the mother, and far safer for the infant.

Besides the above, which may be termed *mechanical obstructions* to labor at full time, there is another class of cases where the induction of premature labor is called for on account of the fœtus alone. There are cases, where, from some diseased condition of the placenta, the child dies in utero before the end of the ninth month. These diseases have been mentioned in previous chapters. There are other cases where the child dies during parturition, owing to a premature detachment of the placenta and consequent hæmorrhage during labor. The remedy for such conditions is premature labor at a time prior to that at which the death of the fœtus has usually occurred.

There is still another class of cases where disease of the mother makes the operation justifiable, — namely, vomiting, chorea, mania, albuminaria, disease of heart and lungs, debility, etc.

Vomiting.— Under the inhuman belief that the life of the child should be saved at the risk of that of the mother, women have been suffered to die from the effects of severe and protracted vomiting, or left to drag out a life of misery with chronic gastritis, or ulceration of the stomach. Several obstetric writers claim that the induction of abortion is never necessary from this cause. Opposed to these, we have the testimony of Dubois, Stoltz, and others, who assert that this is oftener a dangerous occurrence than is supposed. M. Dubois recently stated before the *Academy of Medicine*,* that, in the course of thirteen years, he had met with twenty cases in which vomiting proved fatal to pregnant women. Professor Stoltz of Strasburg relates four cases which came under his observation. In three of these death occurred, and life was saved by the operation in the fourth, although the case seemed hopeless. M. Dubois refers to the fact, that, when the process of gestation becomes arrested, whether spontaneously, or not, vomiting is ordinarily put an end to, although the woman may not be delivered for several days after of a dead child, and may yet die of the effects of what she has already undergone. Women, apparently at the point of death, have been saved by the *spontaneous* death of the fœtus, this being expelled only some time afterward.

There can be no doubt of the moral propriety of terminating gestation in certain cases of vomiting. "The difficulty," says M. Dubois, "is to fix the period at which this should be resorted to; for it is the natural desire to delay this as long as possible, which delay often leads to a fatal result, the woman dying in fact from the exhaustion and prolonged abstinence, which the vomiting has produced, prior to the operation for arresting it being undertaken." Stoltz lays great stress upon the operation being performed in good time, because if we wait until the effects of the sympathetic reaction constitute in themselves a serious disease, the evacuation

---

* Bulletin de l'Acad., tom. xvii., pp. 557–582.

of the womb does not induce a cessation of these, and may, in certain cases, even hasten death, — life, so to say, hanging upon a thread. It is undoubtedly difficult to say *when* the moment has arrived that we can no longer hope for benefit from nature or therapeutical agents. In Homœopathic practice, much may be reasonably expected of the latter; but their use should not be insisted upon too long. M. Dubois lays it down as a rule, "Never to undertake the operation when the signs of extreme exhaustion are present, as evidenced by considerable loss of vision, cephalagia, comatose somnolence, and disorder of the intellectual faculties. On the other hand, we should also abstain from operating when the vomiting, though violent and frequent, still allows of some aliment being retained; when the patient, though wasted and feeble, is not obliged to keep her bed; when the suffering has not yet induced intense and continuous febrile action; and when other means still remain untried. In the first case we would not save our patient, but perhaps accelerate her death, and bring discredit on the operation; while in the other we would sacrifice a pregnancy which might have gone on till the full time. It is therefore the intermediate period which should be chosen, and this is characterized by the following signs:—1. Almost incessant vomiting, by which all alimentary substances, and sometimes the smallest drop of water, are rejected. 2. Wasting and debility, which condemn the patient to absolute rest. 3. Syncope, brought on by the least movement or mental emotion. 4. A marked change in the features. 5. Severe and continuous febrile action. 6. An excessive and penetrating acidity of the breath. 7. The failure of all other means. But even within this period, which is of variable duration, the opportune moment must be chosen. This seems to have arrived when the inefficacy of the most approved treatment has been proved, when fever is found to persist, and the debility and wasting of the patient are making sensible progress. The attendant should now declare that the operation is indicated, leaving the patient and her friends the duty of deciding upon its adoption."

I should consider it an unjustifiable procrastination for the physician to wait until some of the above conditions had obtained. The life of a woman, a wife, a mother, is too precious to be sacrificed to any mistaken sense of duty.

Many *methods of inducing premature labor* have been practised. The earliest plan adopted was that of puncturing the membranes. At first a quill was used, sharpened at the point; afterwards a stillette, or stilletted catheter, was used. This is sometimes described as a canula, with a trocar sheathed in its cavity. The method of operating with this instrument is as follows:

STILLETTE.

Place the patient on her back, near the side of the bed, or better, in the position usually adopted for the operation of "turning," (at right angles with the length of the bed, with the hips near the edge, and the thighs flexed). With the index-finger of the right hand, find the os uteri—the point of the instrument, with the trocar or stilete sheathed, is moved along the palmar surface of the finger until it reaches the os, when it is to be insinuated, up through the canal of the cervix, into the interior of the uterus. The extent to which the point is made to enter the uterine cavity depends altogether upon the period of pregnancy. When it has reached the proper position (against the membranes) the stillette is to be protruded, when it will enter and puncture the membranes. I cannot advise the use of this instrument in any stage of pregnancy, as it is best in all cases to retain the membranes intact as long as possible. The only exception to this rule is, in cases where the membranes are punctured *high up*, in the last months of pregnancy. But in this instance, a *curved* canula is used, and the waters escape very slowly. I prefer, however, other methods to this.

The improper use of the stillette, in the hands of the ignorant and reckless, has caused sudden death. One case came under the observation of Dr. P. H. Hale. A woman procured a stillette of a "nurse," who gave her directions how to use it. She introduced the instrument as far into the uterus as possible, then protruded the sharp point nearly two inches with considerable force. She immediately fainted, and never rallied from the collapse. Death took place in less than half an hour after the injury. There was no flooding. No *post-mortem* was allowed. The husband, who was in the room at the time, did not approve of the operation which the woman attempted to perform. She took the risk, and paid for her criminality with her life. This is only one of many instances where the woman uses such instruments against the will of the husband. This every physician will bear witness to.

In the case above mentioned, How did the stillette cause death so suddenly? It is an interesting question for the pathologist to solve.

But this plan has of late years been discarded as dangerous both to mother and child. It has, however, been revived somewhat by M. Meissner, under another and better form—namely, puncturing the membranes *high up*, by means of a *curved* stilletted catheter carefully introduced between the membranes and the uterus, avoiding the site of the placenta. This plan has been successful in many instances. The waters escape so gradually that the passages become lubricated and relaxed, and pains make their appearance in the course of twenty-four hours.

The *sponge-tent*, invented by Kluge is another plan frequently adopted. It acts both as an irritant and a mechanical dilator, thus bringing on labor by causing reflex action, and opening the os and cervix by its expansion. It is used in the following manner: Warm emollient injections are used; then the conical prepared sponge is carried up to the uterine orifice by means of a pair of long curved forceps (intra-uterine forceps) and is made to enter the canal of the cervix, where it is kept in place by means of a tampon of pieces of cloth, or a sponge. If pains are not established in twenty-four hours, use another and larger tent. Sometimes the plan of M. Meissner is combined with Kluge's with good results.

Small *caoutchouc bags* have been recommended by Dr. Barnes, of England, and used very successfully, instead of the sponge tent. They are introduced into the canal of the cervix, and filled with water. For full particulars relative to this method, reference is made to the articles on that method.

"This instrument," says Dr. Barnes, "is of a fiddle-shape, having, when distended, a narrower cylindrical central portion, dilating at either end into a bulging or mushroom-like expansion. The object of this is to prevent the bag from slipping forward into the uterus, or backward into the vagina. The bag is prolonged into a long narrow tube with a stop-cock at the end, to keep in the water when injected. The injecting medium is the ordinary Higginson's syringe (or 'Essex')—an instrument which should always be carried in the obstetric trosseau, as it is useful for many other purposes. Three bags of different sizes are sufficient as a series. To facilitate the introduction of the flaccid bag into the cervix uteri, a small pouch is attached outside to receive the end of the uterine sound, which, guided by the finger of the left hand applied to the os uteri, serves to push the bag into the cervical canal."

Dr. Barnes claims that by the use of these bags, premature labor may be induced at a predetermined hour. The operation is entirely within the control of the operator. The *vagina* is first dilated by the use of the colpeurynteur; then the smallest or medium bag ("dilator") is introduced into the cervix, care being taken that the terminal bulging part shall pass through the os uteri internum, while the inferior bulging end emerges in the vagina. When water is thrown in, the dilator is thus secured by its shape *in situ*, and the eccentric pressure bears upon the whole cervical canal, and especially upon the two points of greatest resistance—the os externum and internum. This stage ought not, as a rule, to occupy less than three or four hours. If the smallest or medium-sized dilator does not expand the cervix sufficiently, the largest size must be used. If, when full dilation occurs, pains do not set in, a portion of the liquor amnii may be drawn off, and the uterus compelled to collapse. A portion of the liquor should be left in to facilitate turning, if that should become necessary. *Caulophyllin* or *ergot* may be used after full dilation and puncturing of the membranes, if the uterus seems to be in a condition of inertia.

The *tampon*, used as recommended for uterine hæmorrhage, has been resorted to. It should be large enough to expand the upper portion of the vagina, to such an extent as to excite reflex irritation. This plan is, however, painful, and slow in its operation, and is adopted by but few.

The *colpeurynteur* is often used to induce premature labor. The instrument is a simple bag of vulcanized india-rubber, which is to be introduced into the vagina, and then dilated with air, warm or cold water, as may be deemed expedient. After being more or less fully expanded (and it may be advisable at first to gradually inflate it, and increase, as the pressure is less and less inconvenient) the aperture is closed by a stop-cock, and the bag allowed to remain in the vagina. A full description of the instrument and its use was given in the treatment of retroversion. The meaning of the Greek words from which this instrument is derived, is vagina-dilator, and expresses the idea of the effect it produces, dilating the vagina, and even pulling open the os uteri, and by its forcible pressure exciting, by sympathy the action of the uterus.

The *water douche* was introduced into practice by the late Professor Kiwisch, of Wurzburg. Tyler Smith, Cazeaux and others, prefer this method above all others, especially when the life of the child is an object. If a stream of hot or cold water be directed against, or, still better, within the os uteri, at intervals of three or four hours, for the space of ten minutes or a quarter of an hour at each application, labor is certainly and speedily brought on. The water may be made by means of a syphon and reservoir, to descend upon the uterus from a height, or it may be forced into or against the os by a common injecting apparatus which is capable of throwing a continuous stream, like the Essex syringe heretofore described.

Kiwisch explained the *modus operandi* of the douche on the supposition that it caused swelling of the parts by imbibition of fluid, and separation of decidua from the uterus. Probably the reflex and peristaltic actions of the uterus are also excited. Tyler Smith found the douche more efficacious, when warm and cold water were injected alternately, than when either warm or cold was applied alone. The great advantages of the douche are, that the premature labor excited is more certain, and imitates to a greater degree natural parturition; also it never injures the genital organs, the membranes of the ovum, or the fœtus.

Separation of the membranes from the os and cervix uteri, by the catheter, bougie, or sound, is highly recommended by Professor Simpson, who is high authority in obstetric matters. The instrument is oiled and cautiously insinuated between the membranes and the uterine surface, and with short, gentle movements from side to side, carried up towards the fundus of the womb. Care should be taken that the placental vessels are not ruptured. An ordinary flexible bougie, introduced a considerable distance between the membranes and uterus, and allowed to remain, will cause premature labor.

This plan has been improved upon by an eminent obstetrician, who introduces a flexible English catheter, by means of a wire of sufficient size inside, nearly up to the fundus uteri. The wire is then withdrawn, and a quantity of warm olive oil injected into the uterus, *through* the catheter by means of a syringe. Glycerine or milk will answer the purpose as well as oil. Another physician uses a flattened female catheter, to which is attached a common injecting syringe, and after its introduction between the membranes and the uterine wall, a quart of tepid water is injected, slowly, so as to act by gradually separating the membranes.

A syringe, having a very long, slightly flexible tube of hard rubber twelve or fifteen inches in length, would be preferable, as there would be no necessity for the introduction of the catheter. Two openings in the end of the tube, at its sides, would be better than the usual orifice in the end. The amount of fluid thrown up at once should not exceed two or three ounces. Its operation is to separate the membranes over a large surface, and at the same time excite the reflex and peristaltic actions of the uterus.

INTRA-UTERINE SYRINGE.

The *administration of ergot* is recommended by Dr. Ramsbotham, but, as it seems to us, upon insufficient grounds. The majority of obstetricians are opposed to its use, and those who have investigated the action of this drug are the least inclined to use it. If medicines are tried, there are others much safer and more efficient. *Caulophyllin* has in several instances brought on premature labor in the eighth month; the same is asserted of *Cimicifugin*. In order to be effectual, these drugs must be used in large doses — two grains of the former, or one of the latter, repeated every hour or two.

The *application of cupping glasses to the mammæ*, is advised by Scanzoni. Sucking-pumps made of caoutchouc, were applied in one case for two hours, seven times during three days. After the third application the cervix uteri was shortened; after the sixth severe labor pains came on, and after the seventh the child was born.*

Separation of the membranes by the introduction of atmospheric air, or *gas*, has, I believe, been recommended for the purpose of inducing premature labor. I cannot recommend this plan, as I believe it to be decidedly unsafe, and the most dangerous of all methods, not excepting the stiletted canula. Dr. Hitchcock† reports a case which shows conclusively that by this method air may be forced into the uterine sinuses and cause death by entering the circulation.

The case occurred in Kalamazoo county, Michigan, and happened during an attempt to cause abortion. "The operator, through an instrument introduced into the womb, or at least into the vagina, blew air with his mouth, when immediately the woman screamed, struck at the operator, fainted, and was found a minute or two afterwards blue in the face and insensible, the muscles about the arms and neck having a trembling motion." The case presented an entire correspondence with the symptoms described as occurring in cases where death is known to have occurred from the accidental entrance of air into the circulation during surgical operations. The *post-mortem* appearances observed were also altogether consistent with the theory of death from the introduction of air into the circulation; namely, extreme congestion of the lungs, entire absence of blood from the left, and nearly so from the right cavities of the heart; the escape of air from the uterus the instant its walls were incised; the marble whiteness of the extremities, and the unusual paleness of the brain and its membranes.

* Braithwaite, Part XXVIII., page 263.
† Trans. Amer. Med. Association, 1865.

Galvanism. — The employment of *galvanism* or *electro-magnetism* was first suggested by Huder in 1803, for the purpose of bringing on uterine contractions, after all other means failed. This is accomplished by placing one pole of the battery on either side of the uterus, continuing the application of the current for half an hour or an hour each time, and renewing it once or twice daily; the ordinary magneto-electric apparatus in use is the best form, as repeated shocks prove more effectual and certain in stimulating the uterus to contractions than a continued current. In applying the poles, it will be proper to attach to the discs a sponge moistened with water, or salt and water, or pieces of thin flannel likewise moistened may be placed between the discs and the abdomen. Some apply one pole to the neck of the uterus, and the other to the spine or abdomen immediately above the fundus.

Dr. Radford says that " galvanism not only originates the temporary contractions of the uterus, but also produces such a lasting impression on that organ that pains continue to occur until the labor is terminated. It produces severe pains in the loins, and great bearing down efforts, followed by dilation of the os, and expulsive pains.

Dr. King says, " I have employed this agent in a few cases, and with invariable success, though the number and intensity of the applications had necessarily to be varied in each."

In cases of premature labor, the fœtus does not appear to be injured by the application of galvanism. Cases of still-birth seldom occur from its use, while in the majority of cases where *ergot* is used the child is born dead.

Dr. Radford writes, " *Galvanism* is especially advantageous as a general stimulant in all those cases in which the vital powers are extremely depressed from loss of blood. Its beneficial effects are to be observed in the change of countenance, restoring an animated expression; in its influence on the heart and arteries; in changing the character of respiration; and its warming influence upon the general surface. I have several times observed, in cases in which other powerful stimulants have failed to produce any beneficial results, the most decided advantages accrue from its application."

This recommendation would lead us to use this agent in the prostration and chill which sometimes ushers in an abortion. In such cases the abortion is inevitable; and *galvanism* would be productive of great benefit both by bringing on healthy reaction and aiding the uterus in expelling its contents.

Galvanism will cause embryonic and fœtal abortion, as well as premature labor. I have made use of it in several cases. In one it caused an easy abortion in the fourth month (this was a case of dangerous vomiting); in another it acted satisfactorily in the eighth week of pregnancy. In a case where it was decided necessary to terminate the pregnancy, it was used but once, and for five minutes only. Pains immediately set in, accompanied by some hæmorrhage. The abortion would undoubtedly have proceeded to a termination, had not the woman, under the idea of hastening the labor, drank a "hot gin sling." Strange to say, this stimulant immediately arrested the abortivant process, the pain and hæmorrhage ceased, and did not return. (Was it by virtue of the juniper (*sabina*) contained in it?) Four weeks were suffered to elapse, when the abortion was brought on by the use of the catheter.

Dr. Robert Barnes, Lecturer on Midwifery to the

Royal Free Hospital Medical College, England, once wrote a paper on the Use of Galvanism in Obstetrics. It is worthy of attentive perusal, as it enters fully into a consideration of the action of this agent.

The article was published in Braithwaite's Retrospect, Part xxix., pages 259—268, and is copied by Dr. King, in his Obstetrics, pages 669—681.

SECTION II.

FŒTAL ABORTION.

Fœtal abortion consists in the expulsion of the product of conception during the second stage of pregnancy, viz., from the date of the connection of the placenta with the uterus, to the date of *viability* of the child.

The induction of abortion during this stage may be resorted to for the same reasons given above, *where it is not deemed prudent to wait until the child is viable.* The same obstructions, existing to a greater degree, will make it proper to cause the expulsion of the fœtus before the seventh month. In these cases, the life of the mother is to be considered beyond all other considerations.

The same means may be used, and will be found equally efficient as those adopted during the last three months of pregnancy. The same precautions, having in view the non-injury to the placental vessels, are to be adopted, because it is better to have the placenta detached by the contractions of the uterus, than by any instrument. By so doing much hæmorrhage is avoided, which might become dangerous before the uterus could be made to expel the secundines. As before noticed, the hæmorrhage before the sixth month, generally occurs *before* the expulsion of the placenta, whereas, in premature labor, it is *afterwards*.

For the safe and speedy induction of *fœtal abortion*, I prefer, as mentioned under premature labor, *the use of the flexible bougie or catheter, or the injection of bland fluids between the membranes and the uterus*. A combination of the two operations is probably the best that can be adopted. My favorite method is to introduce a smooth flexible catheter or bougie (having one or two orifices in the sides of its extremity) by means of a wire stiff enough to act as a conductor. It should be introduced carefully, so as to gently insinuate it between the membranes and the uterus. I especially recommend the operator not to puncture the membranes, for according to all experience, this accident should be avoided during the whole period of pregnancy. When the instrument has been carried up nearly to the fundus of the uterus, on its posterior surface, the wire is withdrawn, and one or two ounces of warm milk, glycerine, olive oil, or water, is to be injected by means of any syringe which will throw a continuous stream. The *modus operandi* is plain. The fluid permeates between the membranes and the uterus, separating them throughout the whole

FLEXIBLE BOUGIE OR CATHETER.

or a large portion of their extent. It may even separate the placenta during the fourth or fifth month. At the same time the presence of the fluid causes uterine contractions of a forcible character, which generally ends in the expulsion of the fœtus and secundines, with little or no hæmorrhage. The *chill*, or *rigor*, previously mentioned among the symptoms of abortion, is a sure indication that the operation has been successful and efficient.

It sometimes happens that it is difficult, and even impossible, to inject any fluid into the uterus through the catheter or bougie, owing, probably, to the compression of the instrument by the contraction of the cervix uteri, or the whole organ. To avoid this difficulty, I have frequently made use of a hard rubber syringe, holding two ounces, and having a tube about eight to ten inches in length, and about one-fourth or one-sixth of an inch in diameter, with a very smoothly rounded point. (See cut on page 275).

The only instrument of this character now sold, has its tube straight. But before it can be used for the purpose here recommended, it must be bent to the requisite curve. The extent of this curve will depend on the position and size of the uterus, and the period of pregnancy. The tact and judgment of the physician must be the guide in this matter. To prepare the tube so that it can be bent to the requisite curve, it is only necessary to plunge it into boiling water for a minute, when it will become quite ductile, and may be made to assume any curve desired.

When the tube has been introduced into the uterus—as directed for the bougie—the piston is to be slowly pushed down, so as to force the fluid gently out, and effect a gradual separation of the membranes.

In filling the syringe, previous to its introduction, care should be taken to exclude all *air*, as the presence of air in the uterus is neither safe or desirable. The woman should remain in the recumbent position for half an hour or more after the operation, or until the occurrence of the rigors. For obvious reasons, the best time for the operation is in the evening after the patient has retired for the night. It is advisable that the bladder and rectum be emptied of their contents; the latter by enema, if any fœcal matter has accumulated there.

It is taught by some medical writers that it is hazardous to inject fluids into the uterus. Whatever danger there may be in the unimpregnated state from fluids passing through the Fallopian tubes into the peritoneal cavity, thereby causing inflammation, this objection is not valid in the condition under consideration, for during pregnancy those tubes are undoubtedly plugged up with tenacious mucus, which would effectually prevent the passage of any fluid.

This simple and effective method is far preferable to puncturing the membranes, which nearly always results in a tedious labor; or the douche, which is altogether complicated and slow of operation; or the administration of drugs, which is always uncertain, and generally injurious. No other method yet known is at all comparable with it, and in my opinion none other should be adopted by the conscientious physician who seeks in all cases the good of his patient above all else.

SECTION III.

EMBRYONIC ABORTION.

The induction of abortion during the three months following conception, is sometimes necessary and justifiable. If the pregnancy is not to be allowed to go on until such time as the child is viable, it is absurd to wait any time after the physician has decided on the necessity of the operation, as every day's delay renders it more liable to be productive of injury to the woman. The following question was put to Dr. Hunter, in 1768, by W. Cooper, and was shortly after decided in the affirmative by most English practitioners:

"When a woman, three or four months pregnant, has so contracted a pelvis as to preclude all hope of a possible expulsion or extraction of a viable fœtus, may we think of inducing abortion?"

One of the latest obstetric writers—Cazeaux—declares the accoucher is warranted in producing abortion, whenever a woman who is five or six months pregnant, at most, shall have less than two and a half inches in the smallest diameter of the pelvis. I can not see the propriety of waiting until the woman is so far advanced. If the physician is cognizant of the pelvic deformity, and also of the *commencement* of pregnancy, he should induce the abortion at the earliest possible moment consistent with the patient's state of health. He should, of course, be sure the woman is pregnant, before any operation is performed; but it is my opinion that he had better err in this respect, than allow a pregnancy to go on till a late date. For the introduction of the sound or bougie, or the injection of fluids, if carefully done is not productive of injury, and may even bring back the menses, if that was the only obstruction. The indications for producing an abortion during the first *third* of pregnancy, may be thus summed up: Extreme contraction of the pelvis; voluminous, immoveable, and non-operable tumors of the excavation; extreme dropsy of the amnion; irreducible displacements of the womb; hæmorrhages which have resisted the employment of the most rational measures; eclampsia, mania, chorea, and obstinate, dangerous vomiting.

If any of these indications obtain, the sooner after conception the embryo is destroyed, the better for the health and safety of the patient.

Method of Operating.—Nearly all the methods mentioned above, except the injection of fluids, are inoperative when used during the first stage of gestation. They may induce the abortion, but only after a tedious trial, and nearly always to the injury of the uterus. I allude to the sponge-tent, *ergot*, douche, puncture of the membrane, electricity, and colpeurysis.

The most efficient methods are: (1) The introduction of the bougie. (2) The injection of fluids. (3) The forceps. (4) The uterine sound.

The *bougie* or *flexible catheter* is an effective instrument for terminating pregnancy before the end of the third month. If introduced into the uterus without rupturing the membranes, it will bring on labor with expulsion of the ovum entire, with the decidua. Before the fourth month, so close is that tissue to the uterus, that it is a matter of doubt whether a bougie or any other instrument can be insinuated between the two adherent surfaces. The instrument must then pass through the orifice at the cervix, and between the membranes of the ovum and the decidua. It is best not to use sufficient force to rupture the delicate membranes of the ovum, so delicate that I imagine they are destroyed in a large majority of the cases where instruments are used in the early months, especially during the first eight weeks.

To use the bougie successfully, select a smooth one, about one quarter of an inch in diameter, with a curved wire of sufficient size to render it a good conductor (the wires generally found in bougies and catheters are too small for this purpose.) Have the woman lie on her back, near the side of the bed, with the knees drawn up and the hips slightly elevated. First ascertain by the touch the position of the os uteri, its condition, etc. Moisten the instrument with oil or glycerine, and pass its point along the palmar surface of the index finger of the right hand, the end of which should rest upon the lower lip of the os uteri. Insinuate the instrument into the cervix, and push it gently upward towards the uterine cavity until it meets with some obstruction. These obstructions are generally the lacunæ or folds of the mucous lining of the neck. Care should be taken not to lacerate these folds by any forcible measures, as such injuries lead to chronic inflammations of the cervical canal. By the use of tact, patience, and gentle efforts, the instrument will after a time pass suddenly into the uterine cavity. In some cases of primapara, the introduction of any instrument through the cervical canal is a matter of much difficulty. The most careful physician and practical amateur may fail after efforts lasting a quarter of an hour. Sometimes when we are about to give up our efforts, the bougie will glide into the cavity with the utmost ease and readiness. Any one who will take the trouble to examine the minute anatomy of the cervical canal, and note the small size of the inner orifice of the canal, and the numerous lacunæ or crypts of the enlarged middle portion, will understand the reason of the difficulty.

When the bougie shall have entered the cavity of the uterus, it should be slowly pushed upwards until it seems to meet with an obstruction to its further progress. Here all efforts to introduce it further should cease, else we may rupture the membranes of the ovum. Now, with the finger and thumb of the right hand, hold the bougie firmly in its place, and with the other hand slowly withdraw the wire, not directly downwards, but with a circular movement, outward and upward, so that the *curve of the wire will follow the axis of the pelvis*. By so doing we avoid any unnecessary irritation of the interior of the uterus, or a rupture of the membranes.

After the wire has been withdrawn, coil the end of the bougie until it can be placed in the vagina, resting on the perineum, or if it is not flexible enough to be easily coiled, the protruding end may be cut off. At any rate, *leave the instrument in the womb*, until hæmorrhage or labor pains set in, when it may be removed. To keep it *in situ* until such an occasion, no aid is needed if the woman remains in bed ; but if she is to sit up or walk, a bandage should be worn.

The time which intervenes between the introduction of the instrument (if it be retained) and the occurrence of pains, etc., varies from six to twenty-four hours, rarely exceeding the latter. It is very rare that hæmorrhage occurs during a fœtal abortion caused by the proper use of the bougie.

I do not approve of the plan adopted by some obstetricians, of rotating the bougie when in the womb, and then withdrawing it. Not only is injury done thereby, but the induction of abortion by such a method is very uncertain. This leads us to the consideration of

The *Uterine Sound*. Professor Simpson makes use of this instrument for the induction of abortion in the early months of pregnancy. The silver male catheter has been substituted for it in some instances. Professor Simpson directs that after its introduction it is to be turned three or four times round, and then withdrawn. That this method will prove effectual in a majority of cases, I do not doubt; in fact, my observations lead me to believe it will generally destroy the embryo. But Simpson admits that the operation is not always *immediately* effectual, and states that he has been obliged to repeat it several times in some cases, before the uterus would take on expulsive action, or before he was satisfied the embryo was destroyed. There are many other objections to the use of the sound, especially in the third month of gestation. If not used with great care, the interior of the uterus will be abraded or otherwise injured by its forcible rotation. Next to the bougie, however, it is the best and most certain method of causing abortion now known. Its introduction is effected in the same manner as that of the bougie.

UTERINE SOUND.

The *injection of fluids*, as described under *Premature Labor*, may be resorted to when the above means are not at hand, or when it is deemed advisable not to use them. The introduction of the curved tube should be conducted as described above. Not more than one ounce of liquid should be injected at once. When the syringe is used, a sensation of faintness, or actual syncope sometimes occurs immediately after the operation. The rigor occurs sooner than when other instruments are used.

The Forceps.—This instrument is similar to a variety of "bullet forceps," used in military surgery for the seizure and extraction of foreign substances in the tissues. It consists of four "claws," which are sheathed in a tube during its introduction, and are made to protrude and open when in the interior of the uterus. The instrument is pushed upward to the fundus uteri, to the position the ovum is supposed to occupy, when the "claws" are closed upon the embryo, if reached, and the whole forcibly extracted. This instrument is the one most generally employed by those villains who disgrace the medical profession and humanity by practising the vile trade of producing criminal abortion. The instrument may, however, be used by the honorable physician for legitimate purposes. It may be used when the other instruments mentioned are not obtainable, or proper; or for the purpose of removing a retained placenta or membranes, or any residual mass.

ABORTION FORCEPS.

Various other instruments have been used by physicians and abortionists. The *stiletted catheter* was once in general use. There are several varieties of this instrument, but all upon the general plan of a sheathed needle or lancet, which can be protruded when introduced into the uterus, and having for their aim the rupture of the membrane of the ovum.

Besides the *mechanical* measures above enumerated, for the purpose of inducing abortion in the first stage of pregnancy, the *medicinal* means may not pass unnoticed. When premature labor or fœtal abortion is induced by medicines, they set up uterine contractions, and thereby dislodge and expel the uterine contents. But owing to the non-development of the muscular tissue of the womb in the earlier months, such effectual contractions can rarely, if ever, be aroused. Medicines, therefore, which cause abortion during the first three months, must cause it by simulating the menstrual process—namely, *uterine congestion and hæmorrhage, with the exfoliation of the decidua*.

Now, nearly all the medicines mentioned under "Medicinal Causes of Abortion," are capable, under certain circumstances, of causing so much congestion, as to set up a pseudo-menstrual nidus, and the consequent arrest of gestation. If, however, the uterus is in a healthy condition, I do not think such medicines, unless taken in massive and dangerous doses, are ever capable of such effects; but when the uterus is diseased, and has been "irritable," then almost any one of them, if administered at or near the usual menstrual periods, will be very apt to cause such afflux of blood as to lead to the destruction of the product of conception.

SECTION IV.

OVULAR ABORTION.

Another variety of abortion might be considered under the title of *Ovular*.

This might be defined as the destruction of the *ovum* at any time after it leaves the Graafian vesicle, and before it has been impregnated by the seminal fluid.

In other words, any measures which are adopted to prevent the impregnation of the ovum, must result in its destruction.

If, from good and sufficient causes, it is considered best that the fruit of conception should not be allowed to go on to the end of utero-gestation, would it not be better, in a medico-legal, as well as in a moral point of view, to arrest or prevent conception itself?

If premature labor is to be avoided in all cases, and only resorted to to *save the life* of the mother or child, or both; if it is preferable to allow the mother to run the risk of her life, and suffer from dangerous diseases; if fœtal and embryonic abortion is only to be resorted to in the most dangerous cases, and avoided because morally and legally it is a crime, equal to, if not identical with, murder; if so many grave obstacles are in the way, inducing us to avoid any of these operations,—would it not be better, in all cases, for the physician to advise that *ovular abortion* be allowed? By adopting this plan to prevent the occurrence of gestation, no crime is committed, and there is no risk of human life. I admit that this is not feasible in a large proportion of the cases. Physicians are rarely consulted until after pregnancy is well advanced, and often not until the period of confinement itself. But *after* the physician has attended one confinement where the life of the child had to be sacrificed to save the mother, it is his duty to inform his patient of the danger of a future pregnancy, and advise and instruct her as to the best possible manner of *preventing* the occurrence of that state.

All *perfect* theories are probably Utopian. A theory which should regulate and define the laws which relate to marriage is so Utopian that it may never be realized. If it were possible to regulate this matter by absolute law, the following rules would have to be carried into effect:

(*a*) No woman should marry until it was decided by a competent physician that no physical deformity existed which would incapacitate her from bearing a living, average sized child, at full time.

(*b*) After a woman has had one labor, attended with great danger to herself, and involving the death of the child, it should be rendered obligatory upon the parents to prevent the occurrence of future conceptions.

If these two rules were adopted by all civilized nations, it would narrow down the necessity for the induction of abortion to a very small number of cases of difficult and dangerous labor.

Even in the present condition of society, much might be done if physicians would do their duty, and they were upheld by public opinion and public confidence. The fault is both with the physician and people. Physicians are not honest and frank enough with their patients: they do not give them the proper advice and instruction in the premises. We read of women passing through a succession of pregnancies, ending in each case with craniotomy or Cæsarian section, and every time under the care of the same *family physician!* Did the physician do his duty in these cases? Did he instruct the parents how to prevent a conception which was fraught with such dangerous results? Why do not family physicians protest against allowing young women to marry, whom they know, or believe, are not in normal physical condition to bear children? The people are, perhaps, more to blame than physicians, in this matter. This arises from the general ignorance which prevails on all subjects pertaining to the physiology of generation. This ignorance is almost absolute, for not one layman in a thousand, even if he is otherwise well-read and intelligent, has a correct knowledge of this function. In fact, the most ridiculous and absurd notions are held by the people, and, I regret to say, by many physicians, in relation to the passage of the ovum, the manner of its impregnation, and the nature of the seminal fluid.

Books which teach the latest discovered facts relating to generation are kept out of the hands of the young, and virtually out of the hands of the public, by the very nature of the technical language in which they are written. Every man and woman, before entering into marriage, should be conversant, in some way or other, with all the positive facts concerning the generative process.

That I may not be blamed for asserting that physicians, even, are ignorant of facts, I will quote from a work by a distinguished author, who is considered an authority by the profession. Speaking of *impotency*, he says:

"Eacjulation is weak and precipitate, so that the seminal fluid cannot be thrown into the cavity of the uterus * * * it is not sufficient, in order to fecundate, simply to spread the fluid over the vagina: it must be projected with sufficient force through the orifice of the uterine neck."*

This assertion is at variance with all the well-known facts relating to impregnation. A fraction of a drop of the seminal fluid upon the vaginal mucous membrane is sufficient to result in impregnation, if the proper time is selected for the experiment. Not only is the author's knowledge of physiology deficient, but his anatomical knowledge is in a worse condition. The uterine neck is scarcely ever sufficiently open to permit of the seminal fluid being thrown in by ejaculation. It does not open during coition, and need not be open in the manner implied by Lallemand. It may be, and is generally, closed, but not impervious. Nothing short of actual occlusion can prevent the spermatozoa from insinuating themselves into the uterine cavity. The majority of the people entertain an idea similar to Lallemand's, and think they take due precaution, when in fact no real precaution is taken to prevent impregnation.

Ovular Abortion, may be allowed by adopting certain regulations concerning the act of coition.

First, by regulating the time of its occurrence.—In all except about six per cent. of cases, according to M. Rociborski, coition will not result in impregnation, if not performed until ten days after the cessation of the menses, nor within four days previous to, or during their occurrence. Coitus immediately after or during menstruation, has often been advised as a cure for sterility, and frequently with success. Among the Jews,

---

* Lallemand on Spermatorrhœa.

women are not allowed sexual intercourse until twelve days after menstruation; yet the women of that race are noted for their fertility. This is accounted for on the supposition that impregnation took place just previous to menstruation. When conception occurs at this time, the catamenia sometimes appear, and are sometimes absent; if they appear their duration is generally less than usual.*

I am inclined to think that the rates of six or seven per cent. are too small. Some women will conceive at any period between the menses, even when only the slightest particle of seminal fluid is brought in contact with the vaginal surface.

The rule above mentioned, relating to the avoidance of coition at certain times cannot be considered as reliable in all instances. It is based on the theory, which in the majority of cases seems supported by facts, that the ovum is extruded from the generative passages before the twelfth day after, and does not appear therein before the fourth day previous to the menstrual period.

When the experiment has been tried, and found to be reliable, the rule may be adopted to prevent conception, but there is some risk in trying the experiment.

Some women are in the habit, acting probably under the advice of physicians, of taking some powerful emmenagogue just previous to the usual menstrual period. Such drugs may act in two ways — namely, by causing the already impregnated ovum to be expelled with the unnatural menstrual flow, or increasing the amount and force of a natural flow (menorrhagia) which will wash away the unimpregnated ovum in a shorter time than usual.

Second, by regulating the manner of its performance.—

---

* Carpenter, Physiology, page 361.

We have seen that under certain circumstances, in a large proportion of cases, *Ovular Abortion* will occur if coition is not indulged in.

We will now consider the *second* method of inducing ovular abortion; namely, by preventing the seminal fluid from coming in contact with the ovum in the genital passage.

This can only be done by placing some mechanical and impervious impediment in such a manner as will prevent the seminal fluid from escaping into the genital passages.

As the ovum must be impregnated in the uterus, Fallopian tubes, or ovaria, if any obstacle is so placed as to prevent the passage of the spermatozoa through the neck into the cavity of the uterus, impregnation will not take place. As the cervical canal cannot be closed by any mechanical contrivance, we must turn to other means to prevent the entrance of the vivifying animalcules.

Dr. Casanova, in a work more ingenious than reliable,* gives, in a chapter on the "Prophylaxis of Conception," the following as a preventive of impregnation: "The possibility of rendering fecundity ineffectual by artificial means, is founded in the fact that if an appropriate foreign body be interposed between the sexual organs of the male and those of the female, *sub coitu*, fecundation will not take place, because the body will stand as a bar or impediment to the absorption of the *aura spermatica* emitted against it in the act of copulation."

Dr. Casanova illustrates this phenomena by allusion to the discovery of Sir H. Davy, that a wire-gauze lamp prevents the flame within from reaching and

---

* Contributions to Physiology and Medical Jurisprudence, **page 96**.

exploding the explosive gas without. He goes on to say that if a small piece of sponge, or any analogous substance, be placed within the vaginal canal, *sub coitu*, it will produce the same effect as the wire-gauze: *i. e.*, it will represent the intermediary agent which will interrupt the *aura seminalis* from being absorbed from within, thus rendering fecundation impossible."

Although he says, "I have tried the experiment therapeutically more than once, and found the phenomena to correspond, and to be in perfect harmony with truth," we are obliged to deny the truth of his theory, and the reliability of his experiments. There are doubtless cases where this evidently ineffectual means would prevent impregnation, for there are cases where even the slightest precautions are sufficient. In refutation of his assertion, I will state that I have known of many instances where the plan he describes — a plan not new in this country — was carefully and thoroughly tried, and in not one of these instances was it sufficiently effectual to be reliable.

The reason is easily explainable by the fact that only an infinitesimal quantity of the seminal fluid is requisite for the purpose of impregnation. When we remember that the mucous membrane of the vagina is thrown into rugæ or folds, which are more decided immediately after coition, it will be seen that a considerable quantity of the seminal fluid will be caught in these folds, and retained there after the sponge has been withdrawn, for it is impossible for a sponge, unless it be of great size to sweep out all the animalculæ. For this reason also, all those contrivances which have for their object the removal of the semen with a sponge or other similar material, are unreliable in most cases.

Dr. Casanova does not believe that impregnation is

caused by the actual contact of the spermatozoa with the ovum, but by the absorption of what he terms the "*aura seminalis,*" a kind of imponderable force. Even if this were true — and the theory is utterly untenable — it would be so much the worse for his plan, for it is evident that no material substance of a porous nature will prevent the absorption of an imponderable *aura*.

I have been shown an india-rubber ball used for the purpose of preventing impregnation. It was directed to introduce it into the vagina, and place it against the os uteri, and allow it to remain six or twelve hours after coition. This plan, however, afforded no protection from the influence of the semen, for conception took place in spite of it, and for very obvious reasons, namely: the spermatozoa will live several days in the secretions of the vagina. Leewenhoek and other observers have discovered them in a living condition in the uterus and Fallopian tubes, seven and eight days after connection. Of what use, then, is such a contrivance? I have known conception to take place notwithstanding coitus was performed while a large inflated rubber pessary was in the vagina, apparently completely filling up that canal. A study of the nature of the spermatozoa will convince any one that they are capable of insinuating themselves between any such substance and the vaginal wall.

There is but *one* contrivance that is sufficiently reliable to be mentioned and recommended, and one which is absolutely reliable so long as it remains intact, namely: *a covering made to be worn by the male, of sufficient size to cover the whole of the penis.* It should be made of firm, elastic india-rubber, or good gold-beater's skin. The first named material is to be preferred when a good quality is used. They are for sale under the name of "*condom*," by every druggist, and in all pharmacies, and the trade in them is considered legitimate. In selecting them, care must be taken that the material is firm, and contains no small orifices, and will resist the pressure of the air. Some will easily tear when air is forced into them. These are unfit for use, as are those from which the air escapes in minute quantities, with a very low, whistling sound, almost imperceptible to the ear. These pin-holes, as they may be called, are still large enough to permit the passage of a minute quantity of the fluid containing spermatozoa. After each connection, the "covering" should be examined carefully, to see that no rent has occurred during the act, and if such is found to be the case, injections will have to be used, as directed below. Some of these "coverings" are quite durable, and if proper care is taken of them, by cleaning, keeping them dry and inflated, and not allowing them to become adherent to the box where they are placed, they may be used many times, and remain uninjured and impervious.

When it is known that spermatozoa remain in the urethra for many hours after coition, or until free urination occurs,* it will be obvious to all that no second entrance of the penis into the vagina will be safe, unless protected by the covering, or after urination has taken place. Ignorance of the fact has led to the occurrence of impregnation, where the parties were very much astonished at the result.

I will not notice at any length the plan of compressing the urethra so as to prevent the semen from escaping during the orgasm, and until the withdrawal has been effected, for such a plan would be productive of

---

* Lallemand on Spermatorrhœa, p. 262.

the most serious consequences, and result in organic disease of the male organ of generation.

I hardly need allude to the absurd doctrine taught by Dr. Casanova, that unless the woman *participates*, or *enjoys the copulative act*, conception will not take place. His foolish arguments to explain away the facts which disprove this theory, are unworthy the slightest notice from men of intelligence and science. It matters not what the condition of the woman is, whether insensible, indifferent, or absolutely frigid, if the ovum is in the genital passages, and the semen comes in contact with it, impregnation will take place.

We now come to the *third* and last method of causing or permitting Ovular Abortion, namely: *the destruction of the spermatozoa.*

In order to treat this subject in a proper manner, we must make some inquiries into the nature of these peculiar organisms, upon which depend the fertilizing qualities of the seminal fluid.

Nothwithstanding the denial, by Casanova and a few other writers, that the spermatozoa are independent organisms, or veritable animalcules, it is needless, in the present day, to enter into any lengthy refutation of the views of such authorities.

Dr. Casanova says they are not animalcules, "though they possess that optical illusion, a sort of, but not an independent animal life—that is, a life of motion, only, caused by the effects of light."(!) Such an absurd and incoherent assertion is too ridiculous to receive serious notice, and would not be noticed at all in this place, were it not that it is desired to convey to the reader a truthful idea of the nature and importance of these organisms. I am aware that Dr. Carpenter writes that the spermatozoa "have no more claim to a distinct animal character than have the ciliated epithelia of mucous membrane, which likewise continue in movement when separated from the body. They appear to be nothing else than cell-germs, furnished with a peculiar power of movement, by means of which they are enabled to make their way into the situation where they may be received, cherished and developed." These assertions are as untenable as that of Dr. Casanova. I will here mention a few facts in relation to the nature and qualities of the spermatozoa, sufficient to disprove all the above statements.

(*a*) Leewenhock, Gerber, Valentin[*], Dujardin, Wagner, and other eminent microscopists, all testify to having discovered traces of *organization* in spermatozoa. Commenting on this, Hassall properly observes[†] that "the determination of the fact that the spermatozoa are possessed of even the smallest amount of organization, would involve their classification in the animal kingdom."

(*b*) The *motions* of the spermatozoa are proof of their animal nature. Hassall says, "All the spermatozoa contained in a drop of semen which has undergone dilution will not start into motion at once; many of them will remain for a time perfectly motionless, and then suddenly, as it were by an act of volition, begin to move themselves in all directions." Speaking of their "mode of progression," the same writer says,—" The motions of the spermatozoa are effected principally by means of the tail, which is moved alternately from side to side, and during the progression, the head is always in advance." It is stated that the spermatozoa of different animals move in a different manner, because they differ very much in their form and structure. This

---

[*] "The spermatozoa of the bear have a mouth, anus and stomach, or a convoluted intestine. (See illustration in Muller's Embryology, p. 1475).

[†] Microscopic Anatomy, vol. 1, p. 225.

would not be the case if they were "nothing more than ciliated epithelium."

Hassall also states that " in the varied motions executed by the spermatozoa, they exhibit all the characters of volition; thus they move sometimes quickly, at others slowly, alter their course, stop altogether for a time, and again resume their eccentric movements. These movements it is impossible to explain by reference to any hygroscopic properties which may be inherent in the spermatozoa, they appear to be so purely voluntary."

Dr. Morris Wilson[*] says the spermatozoa when moving through a fluid, "*turn readily out of the way of any obstructions*, but they have not the backward motion of vibriones." (Would ciliated epithelium avoid obstructions?)

(c) The *spermatozoa* are influenced by the poisons, or chemical agents, in the same manner as animal organisms. While they "retain their locomotive powers for a very long time in fluids of a bland character—for example, in blood, milk, mucus and pus, on the contrary, in reagents of an opposite character, and in those possessed of poisonous properties, they soon cease to move; thus in saliva and urine, unless these fluids be very much diluted, their motions are soon destroyed, and immediately cease in the acids and alkalies, iodine, strychnine and the watery solution of opium."[†] The narcotic poisons do not arrest the motions of ciliated epithelium.

Lehman[‡] says "the motion of the spermatozoa is destroyed by the solution of opium and strychnine; the tail then generally remains extended." (Is not the ani-

---

[*] "Diseases of the Vesiculæ Seminales," in Lallemand's Spermatorrhœa, p. 346.
[†] Hassall's Microscopic Anatomy.   [‡] Physiological Chemistry, p. 70.

malcule rendered tetanic by the poison?) Hassall thinks, the result of the application of these poisons, furnishes an additional argument in favor of the animality of the spermatozoa, and one which it would be difficult, if not impossible, satisfactorily to controvert.

Dalton* makes the bold assertion that the spermatozoa cannot properly be considered as animals. He says their motions are "precisely analogous" to that of ciliated epithelium. He further says that they are organic forms, produced in the testicles, and forming a part of their tissues, just as the eggs, which are produced in the ovaries, are a part of those organs. Draper† says, "it has never yet been established that anything answering to a true structural arrangement exists, and, upon the whole, it may be concluded that the appearances which have been by some supposed to indicate organization, are, in reality, only an optical illusion."

Notwithstanding the assertions of these high authorities, we must remember that the later discoveries with the microscope have overturned a great many just as bold assertions of the earlier physiologists.

It will be more in accordance with true scientific modes of thought, not to assert of the spermatozoa a want of organization, because we cannot discover and demonstrate it, but to await the results of more minute investigations.

Not only is it necessary to impregnation that the seminal fluid should contain these animalcules, but they must be perfectly organized, and alive.

Wagner found that in the semen of hybrids the spermatozoa were altogether wanting, or occurred in small numbers, and were ill-formed and ill-conditioned. It is

---

* Human Physiology, p. 542. † Human Physiology, p. 519.

a well-known fact that hybrids are incapable of bearing offspring.

In men suffering from impotence, the seminal fluid is found destitute of spermatozoa, or if present, they seem to be either lifeless or of feeble vitality.

It is generally believed that the introduction of semen into the uterus, the spermatozoa of which were dead, would not result in impregnation. M. M. Provost and Dumas, who filtered the seminal fluid, found that the fluid portion, which passed through the filter, would not vivify the eggs (of a frog), while the more solid part, consisting of the spermatozoa, produced impregnation. The apparently contradictory experiment of Spallanzini does not controvert this, for it is *not* absolutely known that the spermatozoa of frogs die in thirty-five hours after being placed in water at a temperature of seventeen to nineteen degrees, or even in fifty-five hours in water three degrees above zero, while it *is* known that the spermatozoa of fishes will live several days and retain their power of impregnation. Moreover, in all cold-blooded animals, cold does not have that destructive effect that it does on the warm-blooded.

From all the above, it is evident that three propositions may be laid down:

I. That impregnation may occur, it is necessary that the semen contain living, perfectly organized spermatozoa, and that these animalcules must come in contact with the ovum while in the living condition.

II. That any agent which is capable, when brought in contact with the spermatozoa, of destroying their life, will prevent their power of impregnation.

III. That if they are wholly washed out of the

vagina and cervical canal immediately after coition, impregnation cannot take place.

It follows, therefore, that to cause *Ovular Abortion*, by other than rules relating to the *time* of the act of coition, or the *mechanical* prevention of the contact of the spermatozoa with the ovum, the results of the second and third proposition must be obtained.

To destroy the life of the spermatozoa while in the vagina, some substance inimical to their vitality must be thrown into that passage in such a manner as to come in contact with them, when holding the seminal fluid in solution.

We must first ascertain the substances which, when thus brought in contact with the spermatozoa, will destroy their vitality and power of impregnation.

Among the most prominent of these agents are, *cold water, watery solution of opium, spirits of wine, salts, acids, alkalies, astringents, strychnine,* and probably all narcotics and other poisons. *Gelseminum, arnica, aconite,* and other medicines, are among the latter. Hassall says the spermatozoa are devoid of life in persons who have died from the poisonous effects of *prussic acid*.

Under certain conditions, any one of the above mentioned agents might be resorted to for the purpose of causing devitalization of the spermatozoa, in order to insure ovular abortion.

The selection of the agent would depend on the health of the woman, and the local diseases to which she was subject. The physician of the parties should be the judge of this matter.

Cold water should generally have the preference in the majority of cases. By reference to a previous paragraph it will be seen that cold water causes motion to cease in human spermatozoa—rolls them up and destroys their power of impregnating the ovum.

Opium would hardly be admissible except under peculiar circumstances. The same may be said of *spirits of wine* and *strychnine*.

The latter, however, may be used cautiously, when there is great relaxation of the vaginal muscles, with prolapsus. I have known vaginal injections of a solution of strychnine, (one grain to one quart of water) to cure a most obstinate case of chronic prolapsus from atony of the muscular tissues. One drachm of the tincture of *nux vomica* in the same amount of water has been equally effectual. Either preparation, in cold water, if injected immediately after coition, would doubtless result in the destruction of the animalcules.

A solution of *common salt* is quite popular as an enema for the prevention of conception. It is used at about the strength of ordinary sea-water. I have been informed by many of my patients who have used this agent, that it cured, in a short time, a profuse and long lasting leucorrhœa, and was quite effectual for the original purpose.

The *acids*, even when used quite dilute, have the effect of destroying the motions of the spermatozoa. *Acetic acid*, or vinegar, largely diluted with water, is used to a considerable extent. *Sulphuric acid* has been recommended, and cases have come under my observation where this was habitually resorted to. The water used as an enema is rendered slightly acid by it, not sufficiently so to cause any smarting of the mucous surface. A very popular and effectual mixture for vaginal enema is made of *sulphuric acid* and *alum*.*

---

* ℞ Acid. Sulph. dil. ʒj.
  Alumen,   ʒj.
  Aqua,   1qt.

Nitric and *muriatic acids* might be beneficial in certain diseased conditions of the vagina and os uteri, and perfectly safe as well as reliable for the purposes desired.

The *alkalies* might be advised or permitted in certain cases. It is said by Donne that when the secretion from the uterus is too alkaline, and that of the vagina too acid, they destroy the life of the spermatozoa and thereby induce sterility.

Astringents have always enjoyed the most extensive popularity. They are resorted to by women in all parts of the world, under the instinctive idea that they contract the mouth of the womb, and prevent the entrance of the seminal fluid. If this was the extent of the action of astringents, their use would be of no value; in fact, they are of but little value unless used in such quantities as to destroy the life of all the animalcules. *Alum* is the most commonly used; *sulphate of zinc* ranks next in popularity. *Tannin*, and vegetable substances containing that substance—namely, geranium, oak-bark, etc., have all been used with success.

Aconite, gelseminum, belladonna, arnica, and *hamamelis,* might each be recommended for the purpose of preventing impregnation. Each would destroy the spermatozoa, and in inflammatory or congested conditions of the os, cervix, and vagina, would prove curative by their local action.

Injections of any medicinal solution, or even of very cold water, no matter how destructive they may be to the spermatozoa, will prove useless unless they are used in a proper manner. It will not suffice to use them in small quantities, or hurriedly. Neither will it do to wait longer than two or three minutes after coition. The spermatozoa are very active when mixed with healthy vaginal mucus; and, in women who have borne children, or in whom the cervical canal is patent, the animalcules enter the os uteri by their own volition in an incredibly short time. They also become lodged in the folds of the vaginal mucous membrane, where they remain unharmed for hours or days if the enema is not most thoroughly used.

The injection, to be effectual for the prevention of impregnation by destroying the spermatozoa, must be used *immediately after* coition; it must be thrown in with considerable force, and in a large quantity, in order to reach the animalcules which may be lodged in the mouth of the womb, or the rugæ of the vagina.

The ordinary vaginal syringe, holding but a few ounces, is not a proper instrument, unless it is very large, and is used six or eight times. The orifices of all vaginal syringes are generally too small, and should be enlarged. The most appropriate instrument is the Essex syringe with an extra large tube. Not less than one pint of water should be used, and if one or two quarts is thrown up, the danger of impregnation is materially lessened in proportion.

The *third* proposition implies the use of *non*-medicated fluids thrown into the vaginal canal for the purpose of washing away all the animalcules lodged therein during coition. For this purpose fluids might be used which were not inimical to the life of the spermatozoa, namely, warm water, milk, etc., and if a sufficient quantity was used would probably be effectual. There would be danger, however, that the bland character of these fluids would preserve the life of a few of the animalcules, and thus allow impregnation to take place.

Cold water is the most generally useful fluid to be used as an enema. It should be used immediately after coition, and in quantity not less than one quart. The temperature need not be lower than 42°, and should not be higher than 60°. An Essex syringe, throwing a continuous stream, should be used, and the water should be directed to all parts of the vagina, particularly the upper portion, around the cervix uteri. The best *position* for the woman to assume is the sitting posture, on the side of a stool or any low seat, in such a position that the water when thrown up will gravitate downward and outward. None of the fluid should be allowed to remain in the vagina.

This last method is preferable to all the others, as a general one. No risk of injury is incurred. I have never known but one instance, where cool vaginal injections could not be borne, and in this it only caused a a temporary spasmodic and painful action of the uterus.

There is but one condition in which it would prove ineffectual; namely, when the uterus lies so low in the pelvis, and the cervical canal so open, as to permit the entrance of the seminal fluid into the uterine cavity during the act of coition. In this state of the organ, impregnation could hardly be prevented, as it is not always safe to inject fluids into the cavity of an unimpregnated uterus.

The following TABLE *will give at a glance the condition of the uterine contents, and the treatment to be adopted, during the* THREE **great** *periods of utero-gestation.*

From conception to end of third month.	From end of third month to end of sixth.	From end of sixth month to full term.
Placenta not developed.	Placenta developed and attached to uterus.	Placenta developed and attached to uterus.
Decidual membrane adherent to the uterus.	Decidual membrane non-adherent.	Decidual membrane non-adherent.
Contractile power of the uterus too feeble to expel embryo, etc., or arrest hæmorrhage.	Contractile power of uterus capable, in some instances, of expelling placenta and embryo.	Uterus capable of forcible expulsive contractions.
Hæmorrhage *before* expulsion of after-birth.	Hæmorrhage *before* expulsion of after-birth.	Hæmorrhage *after* expulsion of after-birth.
REMEDIES FOR HÆMORRHAGE.—*Sabina, erigeron, tanacetum, arnica, hamamelis, trillium,* cinnamon, ice, cold water, hot water, tampon, *sulph. acid.*	REMEDIES FOR HÆMORRHAGE.—*Sabina, crocus, erigeron, erechthites, secale, cinnamon,* ice, and cold or hot water, galvanism tampon.	REMEDIES FOR HÆMORRHAGE.—*Secale, caulophyllin, macrotin, erigeron,* ice, cold and hot water, galvanism (tampon *never*). Pressure on uterus, or manipulations with the hand.
REMOVAL OF DECIDUA.—Blunt hooks—forceps—injections.	REMOVAL OF AFTER-BIRTH.—*Secale,* dry cupping, galvanism, *pulsatilla, caulophyllin, macrotin, gossypium,* reflex irritation by **tampon**, etc.; blunt hook, forceps.	REMOVAL OF AFTER-BIRTH.—Forceps, blunt hook, *secale, caulophyllin, macrotin, gossypium,* galvanism, dry cupping, uterine irritation by cold water or the hands (tampon *never*).

PART VI.

JURISPRUDENCE OF ABORTION.

SECTION I.

The consideration of the Jurisprudence of Abortion, includes the moral and legal bearings of both *obstetric* and *criminal* abortion.

I cannot commence this portion of the work in a more appropriate manner than by quoting entire a Lecture delivered by my venerable friend and colleague, who lately occupied the chair of Medical Jurisprudence in Hahnemann Medical College.

CRIMINAL ABORTION:

A LECTURE BEFORE THE CLASS OF HAHNEMANN MEDICAL COLLEGE, DECEMBER, 1864.

BY PROF. A. E. SMALL.

Next to the crime of infanticide is that of criminal abortion. It matters not by whom committed, whether by the mother herself or some interested friend, nurse or physician. The procuring of abortion, under all circumstances, is a direct violation of the laws of the physical constitution, and almost always a violation of that holy commandment, "Thou shalt not kill."

Before we proceed to discuss the nature of this crime in the light of reason, and in reference to what legislation there has been upon the subject, it is proper to test its heinousness in the light of the *moral law*, which regards the willful killing of a human being, at any stage of its existence, as nothing short of murder. When we consider the fact that fœtal life is human life, distinct from that of the mother's, and dependent upon an organization as distinct from that of the mother's as if it were entirely liberated from its resting place in her womb, we cannot avoid the conclusion that the destruction of such a being would be the destruction of a human life, and that he or she who had an agency in perpetrating the deed would, in the eye of the moral law, be guilty of murder. In order to gain a better understanding of the bearings of the subject, abortion in the abstract must be considered.

To begin with a proper definition—it is a violent and premature expulsion of the product of conception, independently of its age, viability and normal formation. In the investigation of the subject as a crime, all cases of abortion that result from natural causes, or the result of accident, or justified by the rules of medicine, whether to save the life of the mother or her child, will be set aside. We shall confine our discourse to such cases only where **the attempt at premature expulsion of the product of conception is artificially unnecessarily and intentionally made, and without which they would not otherwise have occurred.**

The laws of the land do not recognize that unnecessary abortion *per se* is a crime, inasmuch as the act is **not directed against the life** of the mother, and because, too, she is generally a party to the action performed, and when no manifest injury or loss of life happens to the **mother, it** is regarded a mere misdemeanor; or, if otherwise, the law **does not** take cognizance of the act as a capital offense.

Able authorities upon the subject have pointed out the inconsistency of the law as contemplating the crime as directed against the mother, and not against the fœtus, when in fact no criminal **intent** against the mother can be affirmed, but against the fœtus.

The **act**, when unnecessarily performed, manifestly seems to have been undertaken from one of two reasons—either **to** prevent the product **of** conception from receiving life; or if living, to destroy it. We shall produce evidence to show **that the former cannot be the case,** and, consequently, that the **latter is the sole intent when** the act is committed.

To constitute a crime, a malicious or wicked **attempt is supposed to** exist, and **as** the intent in attempting to **produce abortion is** against the product of conception, and not **against the mother, we cannot but regard this assumption of the law as erroneous; tending** rather to increase the frequent repetition **of the crime** instead **of** exerting a wholesome influence against it. For unless the woman die in consequence of the attempt, it **is** declared, in every state of pregnancy, a mere misdemeanor; **or where** injury is done the mother, not necessarily fatal to her life, the **crime may** be considered a felony, **and** punishable by fine **or** temporary imprisonment. The magnitude of the crime against the second human victim being entirely overlooked.

We shall, for the present, omit **the further** consideration of **the** subject as treated both in common and **statutory law, our** purpose being **to** show that **abortion** is primarily a **crime of the most heinous** character, directed with malicious intent against human embryotic and fœtal life, and in pursuing **this course we** shall attempt to show the fallacy of the arguments **urged by** interested parties in **extenuation of** the offense.

Excepting all accidental and necessary cases **of abortion, it must** be evident that abortions must be intentional, and must **be** occasioned by the "malice aforethought" of the law. It has been stated that the malicious intention, unless otherwise shown, is not directed against the mother, but against the product of her womb. Hence the whole criminality of the offense turns on this one fact—

the real nature of the fœtus in utero. If the fœtus be a lifeless excretion, however soon it might have received life, the offense would have been of minor consequence.

"If the fœtus be already, and from the very outset, a living human being, and existing independently of its mother, though drawing its substance from her, its destruction, in every stage of pregnancy, is MURDER. Every act of procuring abortion," rules Judge King, of Philadelphia, "contrary to the usual interpretation of the law, is murder, whether the person perpetrating such act intended to kill the woman, or merely feloniously to destroy the fruit of her womb."

In Dr. Storer's contributions to Obstetric Jurisprudence may be found a complete reply to the plea of ignorance of the nature of the crime, which is often urged in extenuation. He says:

"Ignorance of the law is held no excuse. The plea of ignorance of guilt could hardly better prevail, where its existence is implied by common sense, by analogy, and by all natural instinct, binding even on brutes. * * * Common sense would lead us to the conclusion that the fœtus is, from the very outset, a living and distinct being. It is alike absurd to suppose identity of bodies and independence of life, or independence of bodies and identity of life; the mother and the child within her, in abstract existence, must be entirely identical from conception to birth, or entirely distinct. Allowing, then, as must be done, that the ovum does not originate in the uterus; that for a time, however slight, during its passage through the Fallopian tube, its connection with the mother is wholly broken; that its subsequent history is one merely of development, its attachment merely for nutrition and shelter—it is not rational to suppose that its total independence, thus once established, becomes again merged into total identity, however temporary; or that life depending on nine months' growth, or on birth, because confessedly existing long before the latter period—since quickening at least, a time varying widely as to limits,—dates from any other period than conception."

"Another argument is furnished us, but differing. The fœtus, previous to quickening, must exist in one of two states, either death or life. The former cannot take place, nor can it ever exist except as a finality. If its signs do not at once manifest themselves, as is generally the case, and the fœtus is retained in utero, it must either become magnified or disintegrated—it can never become vivified. If, therefore, death has not taken place, and we can conceive no other state of the fœtus save one, that—namely, life, must exist from the beginning."

"These reasons are strengthened by the reasonings from analogy. The utter loss of direct influence by the female bird upon its offspring from the time the egg has left her, and the marked effect originally of the male. The independence in body, in movement and in life, of young marsupial mammals, almost from the very moment of their conception, identical analogically with the intra-uterine state of other embryos—nourishment by teat merely replacing that by placenta at an earlier period; the same in birds, shown by movements in their egg, on cold emersion before the end of incubation. The permanence of low vitality or of impaired or distorted nervous force, arising from early arrest or error of development, and necessarily contemporaneous with it, are all instances in point."

"The human instinct, unaided by reason, invariably **leads to the** protection of embryonic and fœtal life. It is said that reason supplants this instinct which is enjoyed in common by the brutes. But this may be doubted, inasmuch as the absence of reason in idiots and **insane** persons does **not** impair the maternal instinct. Whatever ideas the human mind, by reasoning, may have forced itself to believe or entertain, let the slightest proof of the existence of fœtal life be alleged, and maternal instinct **at** once makes itself known."

Thus far, incidental proof concerning the commencement of fœtal life, and consequently the manifest guilt of unjustifiable abortion, becomes apparent. If there exists any doubt of the vitality of **the** contents of the womb in early pregnancy, none whatever can be cherished after the period of quickening.

This period, which betokens the existence **of a** living fœtus in the womb of the mother, declares emphatically that there is an *intrauterine* life that may be destroyed by violen**t** hands; that this positive evidence becomes revealed to the mother by unmistakable sensations, is universally admitted; **and this** accounts **for the** fact of abortions being much more rare after this **period** than **previous.**

But singular **as it** may seem, **quickening is often absent** throughout pregnancy, and other evidences **are relied on to establish the** fact. These cases are confessedly exceptions **to the rule.** In most cases of pregnancy it does occur, but varying **very greatly in** point of time. In the same woman, in successive seasons of pregnancy, the period **of** quickening, reckoning from conception, has varied from fifteen **to** thirty days. In the greater number, the period occurs from the **one** hundred and twentieth to the one hundred and thirtieth day after conception. **But from** *intra-uterine* **causes,** this period may be lengthened to one **hundred** and forty, **or even one** hundred and fifty days, and yet, from **facts** which cannot **be disputed,** the fœtus is living.

We have witnessed premature **births in** several instances where the mother was in doubt of her **situation till** the hour of parturition arrived. Only **a** few weeks ago, a lady **who** had been married two years or more applied for counsel in reference to her health, stated that she had not menstruated for seven months, that her form had changed but little, **and** that at no time had she been conscious of any sensation that would **lead her** to suppose that she had been carrying in her womb a living **child.** At the same time, her feet were swollen, which, in connection with other signs of anasarca, led her to believe that she was the victim of disease not dependent on pregnancy. Scarcely a week, however, elapsed, before there was ocular demonstration of the incorrectness of her conclusions; the hour of parturition arrived, and **she** gave birth to **a** living child.

In premature births, where quickening has not occurred, we have the most positive evidence of early independent and vital existence. These births sometimes occur previous to the time when quickening may be looked for, and in some instances by the ear there is conclusive evidence to be gained of the existence of a living being *in utero*, when no quickening has taken place.

It will be conceded, then, that the period of manifest quickening is by no reasonable conclusion to be regarded the commencement of fœtal life. The period set by some of the old writers was the third day after conception, and this is quite as reasonable, and **yet** neither is to be accredited as true.

In order to arrive at the conclusion that embryotic life exists from the commencement to the end of pregnancy, it is not necessary to set up the claim of sentience and will, as some writers have done. For while perfection of **endowment does not** exist with the embryo, its independent **life can be conceded, with prophetic** endowments for future development, in the **womb and after** birth, of all the **attributes of** human perfection. We must concede this, **if anything,** for how can the embryo merge into **the** fœtus without a perfectly independent excito-motory system, distinct from **that of** the **mother?** If we admit this, we are brought directly to the conclusion, **that** the existence of a distinct and independent nervous centre **must be as** self-acting and living as that of the mother.

If we have succeeded in proving the existence of fœtal life, before quickening has taken or can take place, even though but the beginning of undeveloped faculties,—the inalienable rights of a human being are implied, and this should compel us to believe that unnecessary abortion is a crime.

This once established, we are struck with a full sense of human degradation; mothers imbruing their hands in their infants' blood; fathers **equally guilty,** counseling and procuring the commission of the **crime; nurses** who lend themselves to the infamy; members **of the** medical profession, **made** wicked by their wholesale murders, out-Heroding Herod—eclipsing **in criminality the pirate** upon the seas, the midnight **robber or** highwayman; **wretches who** make this kind of murder **a trade; who go forth with falsehood** and lies, in defiance of the mandates of **the** Almighty; public complacency that nurses the crime, **or** palliates it, or with insulting claims of legal assumption that it is a mere misdemeanor to go forth **as a** destroying angel to prey upon human life. Who can uphold **or defend so** common a violation of all just law, human and divine? **Who** can justify such assault upon all instinct and reason? Let those answer who would seek, if they could, some reasonable excuse for this waywardness.

That criminal abortion is carried on to a great extent in our country, is indeed probable, and that it is carried on apparently by those having high claims to respectability, we also admit, and further, we admit the plausibility of the motives that frequently lead to the act; and these are of sufficient importance to enter concisely into our discussion. Does not the motive give moral quality to the act?—some may inquire. We have already said that necessary and justifiable abortions, in accordance with the rules of medicine, are sometimes demanded. The lives of human beings are, under more circumstances than a few, sacrificed to prevent greater evils. But far be it from us to offer any palliation for the *crime of murder*. When a woman, a wife, either with or without the consent of **her** husband, applies to an abortionist for the purpose of ridding herself of the product of conception, she will always make a show of **reason** for the act, by attempting to show a necessity for it. She may set up the claim of being unable to have the responsibility of children; that she cannot educate them or keep them fed and clothed, **or** above degradation; that her own health will not permit it; that the period of gestation subjects her to confinement and keeps her from society, and she might as well be dead as alive. We have known some physicians to listen to an appeal like this, and yield to **the dictates** of a blind sympathy, especially when a generous fee was **forthcoming** to more effectually close their minds against all sense of moral obligation, and plant themselves upon the flimsy reasoning of such women, and while prostrate before mammon would write a prescription for some powerful deobstruent. But they lend countenance to crime by such acts, and become *particeps criminis* in a murderous transaction.

Such a woman's reasonings are as baseless as a fabric upon the sand. She has a husband to father her offspring, a Heavenly Father to protect her on the voyage of life, and in all probability an angel sent for her future comfort; and would any but a God-forsaken father and mother, and a God-forsaken physician, ever enter into a conspiracy to destroy human life upon such flimsy grounds? Thousands of cases of this kind are constantly occurring **in** the large cities and towns of our country, to enrich such female wretches as Mads. Restell, Beaumont and others, who levy imposing taxes in order to support them in the crime. By all that is human, all **that** is noble and grand in the attributes of true manhood, and by every consideration that relates to time and eternity, every honorable physician should set his face against this practice, and hold it up as murder, in public and private. Let him scorn to sympathize with the barbarity, and feel that fees taken for such agency would produce **an** uncomfortable **sensation** about **his** neck, and bind him as firm as fate to perdition.

But it may be urged that illegitimate pregnancy is a different matter, and that every right-minded physician should be ready to save the reputation of a great many victims of this sort from utter ruin, and their families from disgrace. A mere glance at the crime of abortion, without considering its heinous nature, might cause the unwary physician to fall into temptation. A young lady of previous good reputation has fallen. Will it raise her from degradation to make a violent assault upon the *intra-uterine* life that is the fruit of her imprudence, or, **in** other words, would any physician feel himself worthy of saying his prayers at night, after he had lent his hand to commit a deliberate murder to save the reputation of an already disgraced young woman? What good would he accomplish by the act? Experience proves that the fallen one seldom recovers, and her seducer is made no better, and the physician made more infamous than either.

Such, nevertheless, should command our sympathy, and we should do all we can to promote their temporal, physical and moral well-being. And we can do much in that direction, if we regard first of all our duty to God and our obligation to obey all just laws, both human and divine. It will be proper to guard the reputation as far as we can, and by all possible means consistent with divine and human laws. But God forbid that we should do evil that good may come.

I cordially and sincerely subscribe to most of the views above set forth, so far as relates to the destruction of the ovum, without good and sufficient cause. I differ, however, with some of my professional brethren, in relation to the propriety of inducing abortion or premature labor when certain diseases and conditions exist.

I hold that in no instance should the life, or even *health* of the mother be sacrificed to save that of an impregnated ovum, before the date of its " viability."

The dogma that the embryo, before that date, is of the same importance as after, is yet debatable. I cannot, therefore, look upon the destruction of the ovum before that period as *murder*. If we were to carry the criminality of the deed back to the period of *impregnation*, we may as well carry it back to the period of *ovulation*, and believe with the ancient Hindoos, that the voluntary loss of an unimpregnated ovum was a criminal act. According to Robertson, early marriages in India were obligatory, in consequence of an ancient theory of generation, much resembling the latest modern ovarian theory. It was taught that if an unmarried girl has the menstrual secretion in her father's house, he incurs a guilt equal to the destruction of the fœtus;

for the girl is capable of conceiving, and should be allowed to conceive; menstruation being the loss of an ovum, which is equivalent to the loss of a fœtus.

According to the laws of the Shastras, females must be given in marriage *before* the occurrence of menstruation; and, should consummation of the marriage not take place until after this event, the marriage is a sin.

These ancient doctrines are not to be treated with indifference. The ovum is the receptacle or vehicle of the human soul. If the receptacle is voluntarily allowed to be discharged unimpregnated, it certainly is a criminal offence in a certain degree, for the *existence of a human being is prevented*.

On the contrary, lawgivers or physiologists have no right to establish any date, anterior to viability, after which the unnecessary induction of abortion is a criminal offence.

In relation to this subject, I herewith present a paper on the subject of Criminal Abortion, prepared by my colleague, Dr. Charles Woodhouse, the present incumbent of the chair of Medical Jurisprudence and Insanity in Hahnemann Medical College.

CRIMINAL ABORTION — ITS JURISPRUDENCE.

In our remarks on this branch of our subject, it is a matter of regret that the laws of our country in relation to it, are far from being all that the interests of society demand. But it is the duty of the medical man, and especially the medical author, to do what he can to correct these mistakes, and supply their deficiencies. It is in this conviction that we consider, in this place, this momentous subject.

In medicine, we understand by abortion the expulsion of the fœtus before the sixth month of gestation. If this expulsion takes place between the sixth and ninth month, this, in medical parlance, is termed miscarriage, or premature labor. The popular word for both is miscarriage. And the law does not make the distinction made by medical men. In many instances the law, unfortunately, recognizes the popular falsehood, that a woman may be at a certain point of time, "not quick with child," and in the next instant "quick with child," and this event is supposed always to take place several months after conception. It is needless to say, that no sound physiologist can give the least sanction to this error; an error fruitful in sin and crime; however venerable for age. Lawmakers, **as** well as people, need enlightenment on this matter. Those who believe that at **a** certain period, say the fourth month after **conception, the fœtus for the** the first **moment begins to live, will agree that it is no moral wrong to remove it before that time,** and justify themselves for this crime on this false ground. But a little reflection must satisfy the reason, that the fœtus from the beginning must exist in one of **two** states, *life* or *death*. If dead from the beginning of gestation, that death is a *finality*. The mother **has** no power to give life to a dead fœtus within her. If the fœtus is alive at the fourth or fifth month, or in any period of utero-gestation, it must have been alive from the beginning. Unless the word "quickening," in connection with this subject, can be so used as to guard against error, it had better not be used at all. It is to be hoped that, at all events, the absurd distinction of "quick and not quick with child," will soon be banished from the legal codes of all civilized **lands.** And inasmuch as the laws, in many States, do not, (to use the language of Storer), "recognize the true nature of the crime of **abortion, draw** unwarrantable distinctions, allow many criminals to escape, **neglect to establish a standard of justification,** and in many respects are at variance **with equity and** justice," **every** true physician should aim to perfect, **as far as** his influence **may go,** codes for courts, and ideas for the people, until the Divine **command,** "Thou shalt not kill." shall be understood to apply alike **to** the **taking away** of all human life, fœtal or otherwise.

We will **now give a brief** view of some of the laws of different countries and States, **on abortion**:

ENGLAND.—In this country the absurd distinction of "quick and not quick" is now done away with. Within the reign of the present Queen the following law has been enacted: "Whosoever, with the intent to procure the miscarriage of any woman, shall unlawfully administer to her, or cause to be taken by her, any poison or other noxious thing, or shall unlawfully use any instrument, or other means whatsoever, with the like intent, shall be guilty of felony, and being convicted thereof, shall be liable at the discretion of the court, to be transported beyond the seas for the term of his or her natural life, or for any term not less than fifteen years, or to be imprisoned for any term not exceeding three years." It will be seen at a glance, that this law makes no exception in respect to medical men, who may adopt, for any purpose or with any motive, this practice. It would doubtless be essential for him to show, in case of prosecution for this offence, that his motives were as good as those of the surgeon's in amputating a badly diseased limb, to save himself from the penalties of the law. The English law is clearly against the practice, by any body.

SCOTLAND.—Mr. Allison, a late writer on Scottish law, states it thus: "If a person gives a potion to a woman to procure abortion, and she die in consequence, this will be murder in the person giving it, if the potion was of that powerful kind which evidently puts the woman's life at hazard." Further, "Administering drugs to procure abortion is an offence at common law, and that equally whether the desired effect be produced or not." In 1806 and 1823, persons who used *instruments* for this purpose, were sentenced to transportation for this offence. However it may be in actual practice, the Scotch law, as given by Allison, makes no exception in behalf of medical men. Like the law of England, it shows some advance in a right direction, by discarding the phrase, "quick and not quick," etc.

FRANCE.—The Code of Napoleon declares that every person who, by means of ointments, beverages, medicines, acts of violence, or by other means, shall procure the untimely delivery of a pregnant woman, although with her consent, shall be sentenced to confinement." The mother procuring abortion on herself is similarly punished, under the same code. And it also provides that "Physicians, surgeons, apothecaries, and other officers of health, who shall prescribe or administer such means of abortion, shall, if miscarriage ensue, be sentenced to hard labor for a limited time."

AUSTRIA.—The criminal code of this country, established in 1787, by Joseph II., decrees that a woman with child, using means to procure abortion, is to be imprisoned not less than fifteen years, nor more than thirty, and condemned to the public works. If married, the punishment is still greater. Even advising abortion is severely punished, and where the accomplice is the father of the infant, the punishment is increased.

GERMANY.—Deck gives the following summary of the laws of Germany: If produced within thirty weeks from the time of conception, the woman or her aiders are imprisoned from two to six years. The punishment when committed in the last month of pregnancy, is imprisonment from eight to ten years.

ITALY.—The laws punish the woman who makes the attempt only, to confinement from six months to a year; if she is successful in the attempt, she is confined from one to five years. If the father of the fœtus is a party to the crime, his punishment is still greater. The party attempting abortion against the will of the mother is punished with four to ten years' severe imprisonment; and if the life of the mother is endangered, or her health impaired, the imprisonment is five to ten years.

STATE OF NEW YORK.—The laws of this State, on abortion, are now embraced in three sections, one of which provides that for the willful commission of this crime, by medicine or instrument, the offender shall be imprisoned in the county jail not less than three months, nor over one year. This section omits the distinction of "quick and not quick." Another section provides for punishing the offence, committed on a woman *quick with child*, with intent to destroy the child, unless necessary to save the life of the mother. Otherwise than for this purpose, if either child or mother die in consequence, the offence is manslaughter in the second degree, and the punishment is State prison, for not over seven nor less than four years. A further section provides that the willful killing of an unborn quick child, by an injury to its mother, which would be murder if it resulted in her death, is manslaughter in the first degree, and the penalty is State prison for not less than seven years. This great State has for its motto "Excelsior," and doubtless may for many good reasons still use it; but it needs to follow in the wake of England and Scotland, and younger sister States, in discarding from its laws on abortion the pernicious and absurd phrase, "quick and not quick," etc.

STATE OF CONNECTICUT.—Here this absurd distinction of "quick and not quick," still remains, but the crime may be punished by State prison for life.

STATE OF OHIO.—In the laws of this State, the same absurd distinction is kept up as in New York and Connecticut, but there is a provision justifying the physician in saving the life of the mother, even at the expense of that of the fœtus.

STATE OF MASSACHUSETTS.—In 1845, this State adopted a law, discarding the distinction of "quick," etc., and providing that abortion, by whatsoever means produced, if the woman die, shall be punished by imprisonment for not less than five years, nor more than twenty. If the woman do not die, the offence is a misdemeanor, and the punishment not over seven years, nor less than one, and by a fine of not over $2,000.

STATE OF MISSOURI.—The penalty for this crime here is imprisonment not over seven years, and a fine not over $3,000.

STATE OF VIRGINIA.—This offence is here punished by confinement in the penitentiary not less than one nor over four years.

In all these named States, provisions are made for exempting from punishment any physician, or other person, where the act is done in good faith, with intent to preserve the life of either mother or child. (See Beck. Med. Jurisprudence, Vol. I., page 584).

STATE OF ILLINOIS.—In the laws of this State, the absurd distinction alluded to is not retained. The punishment is confinement in the penitentiary not over three years, and a fine not over $1,000.

STATE OF IOWA.—From the year 1851 to 1858 there were in this State, strange to say, no laws punishing the procurement of abortion. Prior to the year 1851, the "willful killing of an unborn quick child, by drugs or violence," was punished as manslaughter. But for this omission of duty and once legal recognition of the false distinction of "quick and not quick," etc., Iowa has nobly atoned by giving a very good law (a model for some other States) on this subject. It reads as follows: "That every person who shall willfully administer to any pregnant woman any medicine, drug, substance or thing whatever, with the intent thereby to procure the miscarriage of any such woman, unless the same shall be necessary to preserve the life of such woman, shall, upon conviction thereof, be punished by imprisonment in the county jail for a term not exceeding one year, and be fined in the sum not exceeding one thousand dollars."

STATE OF CALIFORNIA.—The law against abortion in California, says *The Medical and Surgical Reporter*, is exceedingly stringent. It declares that the person on whom an abortion is practised shall be held as guilty as the abortionist. The object of this feature of the law is to relieve the physician of vexatious lawsuits, to which he is sometimes subject by the attempt of certain wicked females to fasten upon him criminal abortion, when he has not even been applied to at all in the matter. A case of this kind is said to have occurred in San Francisco, and illustrates the necessity of the law in that State, even though the cases which may occur under it may be rare. Procuring abortion is justly regarded as a crime of great magnitude, and the laws of all the States, so far as we are informed, inflict upon the perpetrator of it a heavy penalty; and the law is right. Its penalties are none too severe; indeed hardly enough so, to prevent its occasional, if not frequent violation.

These citations of some of the laws on abortion, are **sufficient to** show how this crime is viewed in different States and countries, **and** also to call attention to such improvements in them as the facts in physiology **and the** authority **of** the divine **law** shall suggest and require. **The laws** should **in no** case create or perpetuate the distinction **of** "**quick and not quick** with child," as it is repudiated **by right reason and sound physiology.** The fact should also **be pressed upon the attention of** law-makers that a pregnant woman **may, herself, be mistaken as** to whether quickening has taken place **or not, and that** she may bear a living child without having been **ever conscious of** quickening having ever taken place in her **case** at all. The laws should leave no loop-holes for the escape of the offender, **who** knows **the** nature and enormity **of** the crime he or she commits. **One** State, for instance, could be cited, where the laws are perhaps sufficiently severe against the crime of abortion, when produced by *drugs*, **but**, singularly enough, says nothing against **its** production by *instruments*. **The moralist** has a duty to do **in** the premises, and all **books on abortion should be** deemed exceedingly faulty, which do not bear **testimony against this** evil, and help to aid its suppression.

[Dr. Woodhouse **has obtained, with much** difficulty, **the** *existing* laws relating *to* abortion **in the States above mentioned.** In the other States of the Union we cannot ascertain **what** change, if any, has been made in their laws relating to criminal abortion, since the date of the work on that subject by Dr. Storer.]

Dr. Storer, writing in 1860,* gives the following as the laws of the various States:

"**In** the following States—**Rhode Island, New Jersey, Pennsylvania,** Delaware, Maryland, **North** Carolina, South Carolina, **Georgia, Florida,** Kentucky, Tennessee, Iowa,† and the District of **Columbia,** there appear **to** exist no statutes against abortion, and

---

* Criminal Abortion, p. 75. † Iowa has since passed a very good law.

the crime can only be reached at common law and by the rulings of the courts."

So far the instances in this country of an absence of special statutes. Where such exist, they may be variously classified. Reserving for a little all other considerations, we find them at once falling into four great divisions.

1. Those acknowledging the **crime** only after **quickening** has occurred:

  Connecticut,   Minnesota,
  Mississippi,   Oregon.
  Arkansas,

II. Those acknowledging the crime throughout pregnancy, but supposing its guilt to vary with the period to which this has advanced:

  Maine,   Ohio,
  New Hampshire, Michigan,
  New York,   Washington.

III. Those acknowledging the crime throughout pregnancy, unmitigated; but still requiring proof of the existence of this state:

  Vermont,   Missouri,
  Massachusetts,   Alabama,
  Illinois,   Louisiana,
  Wisconsin,   Texas,
  Virginia,   California.
  Kansas,

IV. That, like the present **English** statute, requiring no such proof, and punishing also the **attempt, even though pregnancy do not exist**:

      Indiana.

Briefly to recapitulate these groups:

Maine,	Class II.	Alabama,	Class III.
New Hampshire,	" II.	Mississippi,	" I.
Vermont,	" III.	Louisiana,	" III.
Massachusetts,	" III.	Texas,	" III.
Rhode Island,	no statute.	Ohio,	Class II.
Connecticut,	Class I.	Indiana,	" IV.
New York,	" II.	Illinois,	" III.
New Jersey,	none.	Michigan,	" II.
Pennsylvania,	none.	Kentucky,	none.
Delaware,	none.	Tennessee,	none.
Maryland,	none.	Missouri,	Class III.
District of Columbia,	none.	Arkansas,	" I.
Virginia,	Class III.	Wisconsin,	" III.
North Carolina,	none.	Iowa,	none.
South Carolina,	none.	Minnesota,	Class I.
Georgia,	none.	California,	" III.
Florida,	none.	Oregon,	" I.

MISSISSIPPI.—"The willful killing of an unborn quick child, by any injury to the mother of such child, which would be murder if it resulted in the death of the mother, shall be deemed manslaughter.

"Every person who shall administer to any woman pregnant with a quick child, any medicine, drug, or substance whatever, or shall use or employ any instrument or other means, with intent thereby to destroy such child, unless the same shall have been necessary to preserve the life of such mother, or shall have been advised by a physician to be necessary for such purpose, shall be deemed guilty of manslaughter."* Punishment, by fine not less than one thousand dollars, or imprisonment in the county jail for not more than one year, or in the penitentiary for not less than two years.

ARKANSAS.—"The willful killing of an unborn quick child, by any injury to the mother of such child, which would be murder if it resulted in the death of such mother, shall be adjudged manslaughter.

"Every person who shall administer to any woman pregnant with a quick child, any medicine, drug, or substance whatever, or shall employ any instrument or other means, with intent thereby to destroy such child, and thereby shall cause its death, unless the same shall be necessary to preserve the life of the mother, or shall have been advised by a regular physician to be necessary for such purpose, shall be deemed guilty of manslaughter."†

MINNESOTA.—The willful killing of an unborn infant child, by any injury to the mother of such child, which would be murder if it resulted in the death of such mother, shall be deemed manslaughter in the first degree.

"Every person who shall administer to any woman pregnant with a quick child, any medicine, drug, or substance whatever, or shall use or employ any instrument or other means, with intent thereby to destroy such child, unless the same shall have been necessary to preserve the life of such mother, or shall have been advised by two physicians to be necessary for such purpose, shall, in case the death of such child or of such mother be thereby produced, be deemed guilty of manslaughter in the second degree."‡ Punishment for first degree, imprisonment in the territorial prison for not less than seven years; and for second degree, not more than seven years nor less than four.

MAINE.—Whoever administers to any woman pregnant with child, whether such child is quick or not, any medicine, drug, or other substance, or uses any instrument or other means, unless the same were done as necessary for the preservation of the mother's life, shall be punished, if done with intent to destroy such child, and thereby it was destroyed before birth, by imprisonment not more than five years, or by fine not exceeding one thousand dollars; if done with intent to procure the miscarriage of such woman, by imprisonment

---

* Revised Code of Mississippi, 1857, chap. 64, p. 601.
† Digest of Statutes of Arkansas, 1848, chap. 51, p. 325.
‡ Revised Statutes of Minnesota, 1851, chap. 100, p. 493.

less (sic) than one year, and by fine not exceeding one thousand dollars."*

NEW HAMPSHIRE.—"Every person who shall willfully administer to any pregnant woman any medicine, drug, substance or thing whatever, or shall use or employ any instrument or means whatever, with intent thereby to procure the miscarriage of any such woman, unless the same shall have been necessary to preserve the life of such woman, or shall have been advised by two physicians to be necessary for that purpose, shall, upon conviction, be punished by imprisonment in the county jail not more than one year, or by a fine not exceeding one thousand dollars, or by both such fine and imprisonment, at the discretion of the court.

"Every person who shall administer to any woman pregnant with a quick child, any medicine, drug, or substance whatever, with intent thereby to destroy such child, unless the same shall have been necessary to preserve the life of such woman, or shall have been advised by two physicians to be necessary for such purpose, shall, upon conviction, be punished by fine not exceeding one thousand dollars, and by confinement to hard labor not less than one year nor more than ten years.

"Any person who shall cause the death of any pregnant woman, in the perpetration or attempt to perpetrate either of the crimes mentioned in the two preceding sections, or in consequence of the perpetration or the attempt to perpetrate either of said crimes, shall be taken and deemed to be guilty of murder in the second degree, and be punished accordingly

"Any woman who shall voluntarily submit to the violation of the provisions of this act (this and the three preceding sections†) upon herself, shall be punished by imprisonment in the county jail not exceeding one year, or by fine not exceeding one thousand dollars, or by both said fine and imprisonment, at the discretion of the court."‡

OHIO.—"Any physician, or other person, who shall willfully administer to any pregnant woman, any medicine, drug, substance, or thing whatever, or shall use any instrument, or other means whatever, with intent thereby to procure the miscarriage of any such woman, unless the same shall have been necessary to preserve the life of such woman, or shall have been advised by two physicians to be necessary for that purpose, shall, upon conviction, be punished by imprisonment in the county jail not more than one year, or by fine not exceeding five hundred dollars, or by both such fine and imprisonment.

"Any physician, or other person, who shall administer to any woman, pregnant with a quick child, any medicine, drug, or substance whatever, or shall use or employ any instrument, or other

---
* Revised Statutes of Maine, 1857, chap. 124, p. 685.
† The above should evidently read, "the first two sections," to be possible.
‡ Compiled Statutes of New Hampshire, 1853, chap. 227, p. 543.

means, with intent thereby to destroy such child, unless the same shall have been necessary to preserve the life of such mother, or shall have been advised by two physicians to be necessary for such purpose, shall, in case of the death of such child or mother, in consequence thereof, be deemed guilty of a high misdemeanor, and upon conviction thereof, shall be imprisoned in the penitentiary not more than seven years, nor less than one year."*

MICHIGAN.—"The willful killing of an unborn quick child, by any injury to the mother of such child, which would be murder if it resulted in the death of such mother, shall be deemed manslaughter.

"Every person who shall administer to any woman pregnant with a quick child, any medicine, drug, or substance whatever, or shall use or employ any instrument, or other means, with intent thereby to destroy such child, unless the same shall have been necessary to preserve the life of such mother, or shall have been advised by two physicians to be necessary for such purpose, shall, in case the death of such child or of such mother be thereby produced, be deemed guilty of manslaughter.

"Every person who shall willfully administer to any pregnant woman, any medicine, drug, substance, or thing whatever, or shall employ any instrument, or other means whatever, with intent thereby to procure the miscarriage of any such woman, unless the same shall have been necessary to preserve the life of such woman, or shall have been advised by two physicians to be necessary for that purpose, shall, upon conviction, be punished by imprisonment in a county jail not more than one year, or by a fine not exceeding five hundred dollars, or by both such fine and imprisonment."†

VERMONT.—"Whoever maliciously or without lawful justification, with intent to cause and procure the miscarriage of a woman, then pregnant with child, shall administer to her, prescribe for her, or advise or direct her to take or swallow any poison, drug, medicine, or noxious thing, or shall cause or procure her, with like intent, to take or swallow any poison, drug, medicine, or noxious thing; and whoever maliciously and without lawful justification, shall use any instrument, or means whatever, with the like intent, and every person with the like intent, knowingly aiding and assisting such offenders, shall be deemed guilty of felony, if the woman die in consequence thereof, and shall be imprisoned in the State prison not more than ten years, nor less than five years; and if the woman does not die in consequence thereof, such offenders shall be deemed guilty of a misdemeanor, and shall be punished by imprisonment in the State prison not exceeding three years, nor less than one year, and pay a fine not exceeding two hundred dollars."‡

WISCONSIN.—"Every person who shall administer to any pregnant woman, or prescribe for any such woman, or advise or procure

---

* Revised Statutes of Ohio, 1854, chap. 162, p. 296.
† Compiled Laws of Michigan, 1857, vol. ii. chap. 180, p. 1509. The statute of the Territory of WASHINGTON is very similar to those above.
‡ Compiled Statutes of Vermont, 1850, chap. 108, p. 560.

any such woman to take any medicine, drug, or substance, or thing whatever, or shall use or employ any instrument, or other means whatever, or advise or procure the same to be used, with intent thereby to procure the miscarriage of any such woman, shall, upon conviction, be punished by imprisonment in a coun**ty jail, not more** than one year nor less than three months, or by fine not **exceeding** five hundred dollars, **or** by both fine and imprisonment, **at the discretion of the court.**

"Every woman who shall take any medicine, drug, **substance, or** thing whatever, or who shall **use** or employ **any** instrument**, or shall** submit to any operation, or **other means** whatever, with **intent to** procure a miscarriage, shall, **upon** conviction, be punished by imprisonment in a county jail not more than six months, nor less than one month, or by a fine not exceeding three hundred dollars, **or by** both fine and imprisonment, **at** the discretion of the court."*

ALABAMA.—"Any person who willfully administers to **any pregnant woman, any drug or substance, or uses and employs any** instrument or other means to procure her miscarriage, unless the same is necessary to preserve her life, and done for that purpose, must, on conviction, be fined not more than **five** hundred dollars, and imprisoned not less than **three nor** more than twelve months."†

LOUISIANA.—"Whoever shall feloniously administer, or **cause to** be administered, any drug, potion, **or** any other **thing,** to any woman, for the purpose of procuring a premature delivery, and whoever shall **administer,** or cause to be administered, to any **woman pregnant with** child, any drug, potion, or any **other** thing, for **the purpose of procuring abortion, or a premature delivery, shall be imprisoned at hard** labor for not less than **one, nor more** than ten years."‡

TEXAS.—"**If any person shall designedly administer to a** pregnant woman, with **her consent, any drug or medicine, or** shall use toward her any violence, or any **means whatever,** externally or internally applied, and shall thereby procure **an** abortion, he shall be punished by confinement in the penitentiary not less than two nor **more** than five years; if it be done without her consent, the punishment shall be doubled.

"Any person who furnishes the means for procuring an abortion, knowing the purpose intended, is guilty **as** an accomplice.

"If the means used shall fail to produce **an** abortion, the offender **is nevertheless** guilty of an attempt **to** procure abortion, provided it be shown that such means were calculated to produce that result, and shall receive one-half the punishment prescribed.

"If the death of the mother is occasioned by an abortion so produced, **or** by an attempt to effect the same, it **is** murder.

* Revised Statutes of Wisconsin, 1858, chap. 169, sect. 58. It will be noticed that the second section of the above statute differs from the first, in requiring **the** proof of pregnancy.
 † Code of Alabama, 1852, sect. 3230, p. 582.
 ‡ Revised Statutes of Louisiana, 1856, p. 138. By its wording, this statute might be forced into the next division.

"If any person shall, during the parturition of the mother, destroy the vitality or life of a child, which child would otherwise have been born alive, he shall be punished by confinement in the penitentiary for life, or any period not less than five years, at the discretion of the jury.*

"Nothing contained in this chapter shall be deemed to apply to the case of an abortion procured, or attempted to be procured, by medical advice **for** the purpose of **saving** the life of the mother."†

INDIANA.—"Every person who shall willfully administer to any pregnant woman, or to any woman whom he supposes to be pregnant, anything whatever, or shall employ any means with intent thereby to procure the miscarriage of such woman, unless the same is **necessary** to preserve her life, shall be punished by imprisonment in the county jail **not** exceeding twelve months, and be fined not exceeding five hundred **dollars**."‡

KANSAS.—"**Every physician or** other person who shall willfully **administer to** any pregnant woman, any medicine, drug, **or** substance **whatever, or** shall use or employ any instrument or means whatsoever, **with** intent thereby to procure abortion, or the miscarriage of any such woman, unless the same shall have been necessary to preserve the life of such woman, or shall have been advised by a physi**cian to be** necessary for that purpose, shall, upon conviction, be adjudged guilty of a misdemeanor, and punished by imprisonment in a county jail not exceeding one year, or by fine not exceeding five hundred dollars, or by both such fine and imprisonment." ‖

After commenting on the imperfections of the laws of the States relating to criminal abortion, Dr. Storer makes the following just remarks :

" If our previous assumptions of the actual character of criminal abortion **be granted, and we believe** that they have been proved to a demonstration, **it must follow from the** subsequent remarks that the common law, both **in theory and in practice,** is insufficient **to** control the crime ; that in many **States of this Union the statutory** laws do not recognize its true nature ; **that they draw un**warrantable **dis**tinctions of guilt; that they **are** not sufficiently comprehensive, directly allowing many criminals to escape, permitting unconsummated attempts, and improperly discriminating between the measures employed ; that they require proofs often unnecessary or impossible to afford ; that they neglect to establish a standard of justifi-

---

* **I insert this** clause not merely for its relation to the points we are now **considering, but** for its important bearing on the broad question of infanticide **during labor;** concerning which it stands in bold and direct antagonism to **all the** rulings of the common law in this country and abroad. In other respects also, though not faultless, the Texas statute is rationally and admirably drawn.

† Penal Code of Texas, 1857, p. 103.
‡ Revised Statutes of Indiana, 1852, p. 437.
‖ Statutes of Kansas, 1855, chap. 48, p. 243.

cation, and thereby sanction many clear instances of the crime; that by a system of punishments wholly incommensurate with those inflicted for all other offences whatsoever, they thus encourage instead of prevent its increase; and that in many respects they are at variance, not merely with equity and abstract justice, but with the fundamental principles of law **itself.**

"'It is to be hoped,' has forcibly been written, '**that the period is not far remote, when laws so cruel in their effects, so inconsistent with the progress of knowledge and civilization, and so revolting to the feelings and claims of humanity, will be swept from our statutes.'***

"**In** a similar trust, it now behooves us to consider **whether, and** in what manner, the difficulties in the way of generally suppressing the crime of abortion can be overcome.

CAN IT BE AT ALL CONTROLLED BY LAW?

"To this important question I do not hesitate to give an unqualified answer in the affirmative. The fact that criminal abortion is not controlled by law anywhere, cannot be entertained as a valid argument to the contrary of this assertion; for it is equally the fact, as we have seen, that laws against abortion do not as yet exist, which are in all respects just, sufficient, and not to be evaded.

"It is evident that in aiming to suppress this crime, the law should provide not merely for its punishment, but indirectly as **well as** directly, and so far as possible, for its prevention. The punishment of a crime cannot be just, if the laws have not endeavored to **prevent that crime by the best means which times and circumstances would allow,**† and this is **to be accomplished by a twofold process — by** rendering on the one hand its detection more probable, and on the other its punishment **more certain.**

As indirect though important measures for the former of these ends, we have **already** mentioned laws for registration,‡ and against

---

* LEE, note to Guy's Principles of Forensic Medicine, p. **134.**
† BECCARI, Crimes and Punishments, 104.
‡ "An efficient and practical remedy for the prevention of this crime would be a law requiring the causes of death to be certified by the physician in attendance, or where there has been no physician, by one called in for the purpose. In this way the cause of death, both in infants and mothers, could be traced to attempts to procure abortion. In three cases which occurred in Boston in 1855, the death was reported by friends to be owing to natural causes, and in each it was subsequently ascertained that the patient died in consequence of injuries received in procuring abortion. It is probable that such cases are by no means rare; and if the cause of death were known, an immediate investigation might lead to the detection of the guilty party."— (*Boston Med. and Surg. Journal*, Dec. 1857, p. 365.)

[NOTE.—There is no law of the State of Illinois against the concealment of births, and, we believe, none against secret burials. In this city (Chicago) there is also no ordinance against the concealment of births; and although there is an ordinance providing for the registration of deaths, and against secret burials, yet, owing to the culpable neglect of the city authorities, the concealment of births and secret burials. As a single proof of their possible influence in this respect, out of many that might be adduced, we instance the fact that in Paris the number of premature fœtuses deposited at the Morgue, during the nine years from 1846 to 1854, inclusive, was found to exceed by more than two-thirds that of the full decade just preceding, from 1836 to 1845.* To render this difference more apparent, we have compiled the following table:

Age of Fœtuses deposited.	Ten years: 1836 to 1845.	Nine years: 1846 to 1854.
From 2 to 3 months	21	58
" 3 to 4 "	35	73
" 4 to 5 "	56	102
" 5 to 6 "	69	82
Total	181	315

"Part of this advance, it is true, is attributable to the increase in the population of Paris, and in the prevalence of criminal abortion; but in great measure it is clearly owing to the enforcement of a more rigid law against secret burials. The above remarks are strikingly corroborated by the fact that of trials for the crime—and we must not forget that these bear but a small ratio to the whole number of cases preliminarily investigated†—there were in France, during the latter of these periods, fully four times the number occurring from 1836 to 1845.

"The establishment of foundling hospitals, by the State governments, has been urged as a preventive of the crime, and, on the other hand, fears have been expressed lest the same means should increase it. For ourselves, however, and from some experience in

law is almost a dead letter. Not only do the undertakers neglect to report many cases of burial, but they are unwarrantably permitted to report to the health officer on the causes of death! In consequence the reports of that officer are a standing disgrace to any civilized community, on account of the gross ignorance which they expose. Not only this, but they undoubtedly hide a vast amount of crime, behind their unmeaning and unscientific terms.

We will take as an example, the yearly record of the city mortality for 1865. The whole number of deaths for that year were 3,659. Of this number the following are set down as the causes of death:

Congestion (!!) 90
Cramps (!!!!) 183

What was the matter with these 183 persons, who died of "cramps?" and the 90 who died of "congestion?" Is it not disgraceful that such returns are permitted in a civilized city? How many of these deaths were really from criminal causes? How many resulted from criminal abortion? This is only known to the *All-seeing Eye*, and the guilty ones. There were also 132 still-born infants buried in 1865. Did all these die natural deaths? Does any one suppose that this small number includes all the still-born of this great city?—HALE.]

* Register of the Morgue.

† From 1846 to 1850, 188 cases of criminal abortion were discovered in Paris, but, for want of proof, only 22 of them were sent to trial. (*Comptes Rendus Ann. de la Justice Criminelle.*)

such cases, we believe that these fears are groundless, and that with equal justice might they be entertained of every large charity having for its end the improvement, sanitary or otherwise, of the masses of society.

"We have quoted a statute existing in **Massachusetts, though** practically unenforced, against one great agent in **the increase of** abortion, an abuse of its license by the public press. **Were such** laws to become general, and **to be** faithfully executed, and **were it** also made penal to sell any drug, popularly known as emmenagogue, except as advised by physicians, just **as** the sale of direct poisons is, or should be, controlled by law, **the present** system **of openly** advertising **by abortionists,** would **undoubtedly** be **brought to a close.**

"In no matter is it of more importance than in cases of suspected criminal abortion that coroners should be intelligent **and well educated** medical men; and we could wish **that** this point might **have** received especial attention from Dr. Semmes, in his late admirable report to the American Medical Association.* In the sudden excitement of an inquest, the guilty are more likely than at a later period to be off their guard, and evidence may often be elicited at this time, which, at the subsequent trial, it would be impossible to obtain. There can be no question of the importance of this point; **the coroner should be skilled in all that** pertains to obstetric jurisprudence; and if similar knowledge **were** generally possessed by other officers of justice, attorney, juror, **and** judge, a far greater number of **convictions, under a proper law,** would be secured.

"As regards the **more direct statutes, we have already considered** their important points.

"'In order to render laws effectually preventive,' **as has** wisely been said, '**they should** be consistently framed, **and** based on justice.'† In **accordance with** this truly axiomatic doctrine, and with various rulings **of** the courts, already quoted, no proof should be demanded which is not necessitated by the actual character of the crime. We have **seen** that neither in intent nor in **fact** is this an attempt against the person or life of the mother. If she die in consequence, the offender is already amenable for it **as** homicide; in the absence of any special statute, at common law. **The** crime both in intent and in fact, **is** against the life of **the child.**

"**The** attempt being proved, **it is unnecessary that it should** have been consummated, not merely the completion of a crime bringing its punishment, but also certain overt acts with intent to the perpetration; nor is it requisite that any injury, specific or general, should have been inflicted upon the person of the mother.

"The offence being of equal guilt throughout pregnancy, proof of quickening, the incident, not the inception of vitality,—indicating neither the **commencement of a** new stage of existence, nor an

---

* Report on the Medico-legal Duties of Coroner. 1857.
† RADFORD, British Record of Obstetric Medicine, vol. i. p. 55.

advance from one stage to another,* —and, therefore, an element without the slightest intrinsic value, should not be required.

"The crime of abortion should be considered to include, as it does, in the absolute fact of moral guilt, **all cases of** attempted or intentionally effected destruction and miscarriage of the product of impregnation; and this, whether it be living or dead, normal or abnormal, which last expression equally comprehends instances of moles, hydatids, extra-uterine conception, acephalous, anencephalous, and other monsters.

"Proof should not, as now, be required of **intent to destroy the child.**† This should be considered shown by **the intent** to produce miscarriage, in the absence of lawful justification therefor; the act in all stages of pregnancy being attended with great danger to the child, and, in much more than a moiety of the period, necessarily fatal to it.

"The attempt being considered criminal, it follows that proof of pregnancy is not necessary, and that conviction should be had though it were proved that pregnancy did not exist,‡ even that the woman on whom the abortion was attempted, however unlikely, was still a virgin.‖

"No discrimination should be made as to the means criminally employed, and no escape thus afforded to the guilty; as we have seen still obtaining in Great Britain and many of our own States.

"The mother, almost always "an accessory before the fact," or the principal, should not, as now, be allowed almost perfect impunity. There is no valid **reason for** such exemption; there is every reason against it. The woman is covered by the laws of most continental nations of Europe—France, Austria, Germany, Bavaria, and Italy,—and by many of them her punishment, if married, is greatly increased. Similar severity is also exercised in these countries against the father of the fœtus, if he, too, is implicated in the crime.

"To allow that abortion is extenuated in the unmarried, it has been said, will 'to the moral and political philosopher appear to have exalted the sense of shame into the principle of virtue, and to have mistaken the great end of penal law, which is not vengeance, but the prevention of crime. Law, which is the guardian and bulwark of the public weal, must maintain a steady and even rigid watch over the general tendencies of human actions."§ But, on the other hand, the measure of punishment should be proportionate, as nearly as possible, to the temptation to offend, and to the kind and degree of evil produced by the offence."¶

"We have seen the increase in moral guilt, and of opportunity for commission and for escape, in the case of nurses, midwives, and

---

* Wharton, Criminal Law, 540.
† Smith vs. The State, 33 Maine, (3 Red.) 48.
‡ Rex vs. Phillips; Regina vs. Goodall; Regina vs. Heynes, etc.
‖ Taylor, Medical Jurisprudence, p. 386.
§ Percival, Medical Ethics, p. 84.
¶ Ibid, p. 85.

other classes of persons, who, from their profession, are brought **more** directly into contact with pregnant women. By the penal code of Napoleon the First, remarkable in so many respects for the wisdom of its provisions, an increase of punishment was enacted for abortion criminally induced or advised by physicians, surgeons, or other officers of health, including midwives, or by druggists;[*] their guilt being enhanced by their greater opportunities and knowledge.

" Punishments for the crime of abortion should not, as is now generally the case, be **so** framed as **to** render the statute, in fact, if not in name, simply nugatory. Were the murder of adults to be made answerable by merely a year or two in prison, far more convictions **than at present** would undoubtedly be secured; but it is certain that the instances of the crime would be fearfully increased. We have **reason** to believe **that it is** precisely thus with the case in hand.

" A standard of justification for the instances of necessary **abortion** should be fixed by law. If perfection in this respect be impossible, let the nearest approach be made to it that can. Since my remarks upon the relative rights of the mother and fœtus, to the chance of life in doubtful cases, were published in a former paper of the present series, I have received from Dr. Rattermann, late of Tübingen, an essay, written by himself, in which this question is discussed at length, and the repetition of **abortion upon the same** individual, in the early months of **pregnancy, is defended.** I have carefully considered the several **arguments advanced by the gentleman,** and am compelled **to adhere to the views I have already** expressed.

"In presenting a report upon **the matter, in** 1857, **by** direction of the Suffolk District Medical Society of Massachusetts, the writer offered the draft of a law, prepared after much thought and consultation, **with** legal **as** well as with medical men, and embodying the suggestions made above. **This was** intended for the consideration of the Legislature of the **State, in the** hope that it might be of aid toward a modification of the present defective law.

" Having seen no reason to change the opinion then avowed, but on the contrary, receiving constant confirmation of their truth, I now present the essential portions of that draft, acknowledging most willingly that its wording may, perhaps, with safety, be simplified and condensed; but contending, in all sincerity and earnestness of purpose, that its general tenor is what justice and humanity alike, and imperatively, demand at the hands of society.

"' Whoever, with intent to cause and procure the miscarriage **of** a woman, shall sell, give, or administer to her, prescribe for her, or advise, or direct, or cause, or procure her to take any medicine, or drug, or substance whatever, or shall use, or employ, or advise any instrument, or other means whatever, with the like intent, **unless the** same shall have been necessary to preserve the life of such

---

[*] Loc. cit., article 317.

woman, or of her unborn child, and shall have been so pronounced (in consultation) by two competent physicians; and any person, with the like intent, knowingly aiding and assisting such offender or offenders, shall be deemed guilty of felony,' etc. etc.; 'and if such offence shall have been committed by a physician, or surgeon, or person claiming to be such, or by a midwife, nurse, or druggist, such punishment may be increased at the discretion of the court.

"'Every woman who shall solicit, purchase or obtain of any person, or in any other way procure, or receive, any medicine, drug, or substance whatever, and shall take the same, or shall submit to any operation or other means whatever, or shall commit any operation or violence upon herself, with intent thereby to procure a miscarriage, unless the same shall have been by two competent physicians (in consultation) pronounced necessary to preserve her own life, or that of her unborn child, shall be deemed guilty,' etc. etc.; 'and if said offender be a married woman, the punishment may be increased at the discretion of the court.'

"It was also advised that the encouragement of criminal abortion, by publication, lecture or otherwise, or by the advertisement, sale, or circulation of such publication, should be made penal, and that the present well-worded statute against the personal advertisements of abortionists, and their nostrums, should be rigorously enforced.

"To the words now quoted were added, and they are still applicable, the following:

"'We have aimed at a statute, which, while it better defined this atrocious crime, and covered the usual grounds of escape from conviction, established also the proper standard of competence in all medical questions involving issues of life and death. We believe that it would be the means of preventing much of the present awful waste of human life. But enforce such a law, and the profession would never allow its then high place in the community to be unworthily degraded; nor, as now, would those be permitted, unchallenged, to remain in fellowship, who were generally believed guilty, or suspected even of this crime.'*

* Report to Suffolk District Med. Society, May, 1857, p. 12.

GENERAL INDEX.

Abortion, statistics of, 19
" causes of, 31
" treatment of, 153
" of animals, 108
" forceps, 288
" vectis, 245
Actual symptoms of abort'n, 124
Actea alba, a cause of " 92
Acids in treatment of anæmia, 160
Acetic acid as a preventive of conception, 305
Aconite in treatment of abortion, 220
Aconite as a preventive of conception, 306
Adhesion, treatment of, 253
Aloes, a cause of abortion, 92
Aletris farinosa a cause of abortion, 93
Aletris farinosa in treatment of abortion, 222
Alum as a preventive of conception, 305
Alkalies as a preventive of conception, 305
Alabama, laws of, on abort'n, 330
Anæmia, a cause of abortion, 35
" treatment of, 154
Anteversion, a cause of abortion, 84
Anteversion, treatment of, 214
Appearance on examination of cervix uteri, 83
Apis mellifica, a cause of abortion, 91
Arnica in treatment of abortion, 218

Asarum europeum, a cause of abortion, 92
Asarum canadense, a cause of abortion, 93
Asclepias incarnata, a cause of abortion, 93
Asclepias syriaca, a cause of abortion, 93
Astringents as preventives of conception, 306
Atrophy of placenta, a cause of abortion, 44

Baptisia tinctoria, a cause of abortion, 94
Belladonna in treatment of abortion, 220
Belladonna as a preventive of conception, 306
Blows, Treatment of, 189
Blunt hook, 229
Borax, a cause of abortion, 94
Bovista, a " " 95

Calcareous degeneration of placenta, a cause of abortion, 44
Calcareous degeneration of placenta, treatment of, 175
California, laws of, on abortion, 324
Caoutchouc bags, 271
Can abortion be controlled by law, 332

340 GENERAL INDEX.

Case of retroversion of the uterus, 257
Causes of abortion, 33
Cancerous ulcer, a cause of **abortion**, 80
Cantharis, a cause of abortion, 96
Caulophyllin, a cause of ab'n, 96
" in treatment of abortion, 220
Cerebro-spinal meningitis, a cause of abortion, 49
Centric causes of abortion, 45
" causes, treatment of, 176
Cervical leucorrhœa, a cause of abortion, 44
Chancre simplex, **treatment of,** 208
Cholera as a cause of abortion, 39
Cholera, treatment of, 170
Chlorosis, treatment of, 159
Cimicifuga racemosa, a cause of abortion, 97
Cimicifuga racemosa in treatment of abortion, 220
Cinnamon in treatment of hæmorrhage, 221
Classification of States according to laws, 326
Cold **water** in treatment **of** abortion, 226
Common **salt as a preventive of** conception, 305
Complications **occurring with** abortion, 131
Congestion of the placenta, a cause of abortion, 43
Congestion of the uterus, a cause of abortion, 61
Congestion of **the uterus, treatment of,** 191
Congestion of the ovaries, a cause of abortion, 90
Concentric causes of abortion, 49
Concentric causes, treatment of, 178
Coitus as **a cause of abortion, 87**
Corroding ulcer, a cause of abortion, 80
Corpulence, treatment of, 157
Constipation, treatment of, **186**

Colpeurynteur in treatment of abortion, 226
Colpeurynteur **to produce abortion,** 273
Condom, description **and use of,** 298
Concealment of births and **deaths,** 332
Conduct of physicians when called to case of abortion, 234
Connecticut, laws of, on abortion, 323
Criminal abortion considered, 313
" " is it increasing, 18
" " its jurisprudence, 320
Cystitis, treatment of, 186

Danger from use of stillette, 270
Date of viability of fœtus, 117
Death of the embryo, a cause of abortion, 87
Decidua, character of, 116
Decodon verticillatus, a cause of abortion, 99
Definition of abortion, 36
Dental causes of abortion, 56
" treatment of, 180
Diagnosis of abortion, 133
" " after fourth month, 139
Diarrhœa, treatment of, 185
Diameters of child's head, 265
" pelvis, 266
Dimensions of fœtus at different periods, 119
Dietetic treatm't of abortion, 251
Direct blows upon brain, treatment of, 177
Diphtheria, treatment of, 168
Discharges after abortion, 131
Displacements of the uterus, a cause of abortion, 84
Douche, in treatment of abortion, 273
Dropsy of the ovaries, a cause of abortion, 90

GENERAL INDEX. 341

Dropsy of the ovaries, treatment of, 254
Dry cupping in treatment of abortion, 233
Draft of abortion law, by Doctor Storer, 336
Duty of physician concerning prevention of conception, 291
Dysmenorrhœa, diagnosis of, 135
Dysentery " 135
" treatment of, 183

Embryonic abortion, 283
Emotional causes of abortion, 45
" treatment of, 176
Engorgement of the cervix uteri, 192
England, laws of, on abortion, 321
Epilepsy, treatment of, 189
Ergot, as a cause of abortion, 47
Essex syringe, 203
Exanthematous fevers causing abortion, 38

Falls, treatment of, 189
Fatty deterioration of chorion and placenta, a cause of abortion, 41
Fatty deterioration of chorion and placenta, treatment of, 174
Fecundity in European countries, 20
Fœtus, when so named, 121
" date of viability of, 117
" weight and length of, at different periods, 119
Fœtal abortion, when necessary, 279
Ferrum in Anæmia, 160
Fevers, yellow, 39
" exanthematous, 38
Fissured ulcer, a cause of abortion, 77

Fissured ulcer, treatment of, 205
Fissure of the anus, 185
Fistulas, treatment of, 253
First stage of labor, management of, 244
Flexible bougie, manipulations with, 280
Flexible catheter, manipulations with, 284
Follicular ulcer, a cause of abortion, 79
Follicular ulcer, treatment of, 206
Forceps, long, 231
" short, 231
French laws on abortion, 322
Frequency of " 22
Fucus vesiculosis in corpulence, 158
Functional diseases of uterus, causes of abortion, 61
Functional diseases of uterus, treatment of, 191

Galvanism, a cause of abortion, 48
" in treatm't of " 233
" to induce " 277
Gastric irritation, a cause of abortion, 56
Gastric irritation, treatm't of, 182
Gelseminum in treatment of abortion, 221
Gelseminum to prevent conception, 306
Generation, 115
Germany, laws of, on abortion, 322
Gonorrhœa, a cause of " 69
" treatment of, 197
Granular vaginitis, treatment of, 188

Habitual abortion, 59
Hæmorrhage, diagnosis from abortion, 137
Hæmorrhage, after sixth month, treatment of, 239

Hæmorrhoids, treatment of, 185
Hot water in treatment of abortion, 227
How to use stillette, 269
" destroy spermatozoa, 299
" render laws efficient, 334
Hydatids, a cause of abortion, 41
" treatment of, 175
Hydrorrhœa, diagnosis from abortion, 137
Hypertrophy of placenta, a cause of abortion, 44
Hypertrophy of the uterus, treatment of, 253
Hypertrophy of cervix, symptoms of, 183
Hypertrophy of cervix, treatment of, 192
Hysteria, treatment of, 189

Ice in treatment of abortion, 226
Illinois, laws of, on " 324
Ilex opaca, a cause of " 100
Impregnation, 293
Indiana, laws of, on " 331
Induration of the cervix, a cause of abortion, 82
Indurated chancre, treatment of, 208
Inflammation of placenta, a cause of abortion, 43
Inflammation of ovaries, a cause of abortion, 90
Inflammation of mammæ, treatment of, 253
Injections, 288
" when to use, 307
Instrumental irritation, a cause of abortion, 88
Intra uterine syringe, 275
Iowa, laws of, on abortion, 324
Irregularity of os uteri, treatment of, 200
Italy, laws of, on abortion, 323

Jewish customs, 294

Jumping, treatment of injuries from, 189
Jurisprudence of abortion, 320

Kansas, laws of, on abortion, 331

Laudanum in treatment of abortion, 222
Laws of the Shastras on abortion, 320
Laws of foreign countries on abortion, 321
Laws of different States on abortion, 323
Lecture on criminal abortion—Prof. A. E. Small, 313
Lecture on jurisprudence of abortion—Prof. C. Woodhouse, 320
Length of fœtus at three to seven weeks, 119
Length of fœtus at two to five months, 120
Length of fœtus at six months to full term, 121
Leucorrhœa, a cause of abortion, 62
Leucorrhœa, cervical, a cause of abortion, 64
Leucorrhœa, vaginal, a cause of abortion, 67
Leucorrhœa, treatment of, 192
Leucorrhœa, cervical, treatment of, 192
Leucorrhœa, vaginal, treatment of, 194
Leucorrhœa, sequelæ, treatment of, 65
Local causes of abortion, 39
" diseases, treatment of, 171
" syphilis, symptoms of, 81
Louisiana, laws of, on abortion, 330

Maine, laws of, on abortion, 327
Malformation of the ovum, 172

Malformation of the membranes, 172
Mammary irritation, a cause of abortion, 56
Mammary irritation, treatment of, 179
Management of labor, 241
Manner of coition to prevent conception, 294
Manchester Lying-in Hospital, statistics of, 23
Massachusetts, statistics of, on abortion, 22
Massachusetts, laws of on abortion, 324
Means employed to destroy spermatozoa, 304
Mechanical abortion, pathology of, 142
Mechanical obstruction, 265
" treatment of sequelæ of abortion, 256
Medicinal causes of abortion, 46
" " treatment of, 178
" treatm't of abort'n, 217
" " in the third stage, 252
Medicines acting as predisponents to abortion, 91
Medicines acting as centric causes of abortion, 91
Medicines acting as concentric causes of abortion, 91
Membranes, how separated from uterus, 274
Membranes, how separated by air, 276
Menstrual crisis, a cause of abortion, 36
Menstrual crisis, treatm't of, 162
Mental aberrations, treatment of, 254
Menorrhagia, chronic, treatment of, 255
Mercurialization, a cause of abortion, 38
Mercurialization, treatm't of, 164
Mercury, a cause of abort'n, 101
Methods of inducing premature labor, 269

Methods of inducing embryonic abortion, 284
Methods of using syringe, 308
Metritis, diagnosis from abortion, 134
Metritis, treatment of, 200
" puerperal, treatment of, 254
Missouri, laws of, on abort'n, 324
Minnesota, laws of, on " 327
Michigan, laws of, on " 329
Moles as causes of abortion, 40
" treatment of, 175
Muriatic acid, to destroy spermatozoa, 306

Neuralgia of ovaries, treatment of, 312
New Hampshire, laws of, on abortion, 328
New York, laws of, for abortion, 323
New York, statistics of abor'n, 20
" " " from 1804 to 1862, 21
New York, ratio in, 21
Nitric acid, to prevent conception, 306
Non-medical fluids, use of, 307
Nux vomica, to prevent conception, 305

Obstetric abortion, 263
Obesity, a cause of abortion, 35
Occurrence of abortion, tabular view, 26
Ohio, laws on abortion, 323
Opium, to prevent concept'n, 305
Organic diseases of uterus, a cause of abortion, 71
Organic diseases of placenta, treatment of, 173
Ovarian causes of abortion:
Irritation, 35

Congestion, 58
Inflammation, 90
Tumors, 90
Dropsy, 90
Ovarian diseases, treatm't of, 211
" irritation, " 211
" tumors, " 212
Ovaritis, " 212
Ovaralgia, " 212
Ovular abortion, " 290
" " when necessary, 290
Ovum, 115

Paris, statistics of, on abort'n, 20
Pathology of " 142
Parotidean irritation, a cause of abortion, 55
Parotidean irritation, **treatment** of, 178
Paralysis, treatment of, 256
Patient, examination of, 237
" position of, 243
Pelvic diameters, 266
" cellulitis, treatm't **of**, 253
Peritonitis, diagnosis **from** abortion, 135
Period of pregnancy at which abortion is most frequent, 25
Period of occurrence of abortion, tabular view, 26
Period of abortion, tabular view, 309
Pessaries, a cause of abortion, 88
" ring, 259
Podophyllum, a cause of abort'n, 102
Position of patient **during** injections, 308
Postural treatm't of abort'n, **248**
Post partum " " 248
Phagædenic ulcer, a cause of abortion, 80
Phagædenic ulcer, treatment of, 206
Phlebitis, treatment of, 254
Physician's duty when applied to to produce abortion, 318

Physical causes of abortion, 45
Physiology of generation, 115
Plethora, a cause of abortion, 35
" treatment of, 155
Placental causes of abortion:
Fatty degeneration, 42
Congestion, 43
Inflammation, 43
Calcareous degeneration, **44**
Tubercular deposits, 44
Atrophy, 44
Hypertrophy, **44**
Previa, 44
Placental fatty degeneration, treatment of, 174
Placental calcareous degeneration, treatment of, 175
Placenta previa, treatment of, 172
" retention of, 140
Predisposing causes of abortion, 35
" treatment of, **155**
Pregnancy, signs of, 133
Premature labor, **263**
" when necessary, 264
Premonitory symptoms of abortion, 133
Preventive treatment of abortion, 155
Principal medicines as causes of abortion, 91
Prognosis of abortion, 148
" when favorable, 148
" " unfavorable, 150
Prolapsus, a cause of abortion, 84
" treatment of, 213
" mechanical treatment of, 256
Puerperal **Peritonitis, treatment** of, 254
Puerperal Metritis, treatment of, 254
Pulsatilla in **treatment of** abortion, 222

Quinine, a cause of abortion, 46

GENERAL INDEX.

Quinine sulphas, a cause of abortion, 103

Ratio of abortion in N. York, 21
Rectal irritation, a cause of abortion, 57
Rectal irritation, treatm't of, 183
Reflex causes of abortion, 44
" " treatment of, 176
Regulation of coition, 294
" " Jewish custom, 294
Register at the Morgue, 333
Remote consequences of abortion, 151
Remedial treatment of abortion, 217
Renal causes of abortion, 57
Results of abortion, 28
Resumè of treatment, 209
Retention of ovum, a cause of abortion, 140
Return of menses, a cause of abortion, 36
Return of menses, treatment of, 162
Retroversion, a cause of abortion, 84
" treatment of, 214
" mechanical treatment of, 256
Rules for use of blunt hook, 228
Ruta graveolens, a cause of abortion, 103
Ruta graveolens, treatment of, 221

Sabina, a cause of abortion, 104
" in treatment of " 221
Sanguinaria, a cause of ab'n, 109
Sarracenia purp., 166
Scarlatina, treatment of, 166

Scotland, laws of, on abort'n, 322
Scrofulous diathesis, a cause of abortion, 36
Scrofulous diathesis, treatment of, 162
Secale corn., a cause of ab'n, 106
" " in treatm't of " 221
Separation of membranes from uterus, 274
Separation of membranes from uterus by air, 276
Sequelæ of leucorrhœa, a cause of abortion, 65
Sequelæ of abortion, 131
" " 248
Shastras, laws of, on abort'n, 320
Simple irritation of ovaries, a cause of abortion, 89
Simple granulating ulcer, treatment of, 204
Softening of the uterus, 206
Sponge tent, 271
" in treatment of abor'n, 225
Spermatozoa, character of, 298
" how destroyed, 299
Spirits of wine, to prevent conception, 305
Stage of pregnancy, first, 244
" " " sequelæ of, 249
Stage of pregnancy, second, 246
" " " sequelæ of, 249
Stage of pregnancy, third, 247
" " " sequelæ of, 250
Statistics of abortion, 19
in foreign countries, 20
in New York, 20
in Boston, 22
in Massachusetts, 22
in Chicago, 22
Manchester Lying-in Hospital, 24
at the Morgue, 28
States, laws of, on abortion, 323
which have no laws on abortion, 325
which acknowledge the crime after quickening, 326

GENERAL INDEX.

which acknowledge the crime, but require proof of pregnancy, 326
which punish an attempt at abortion, **326**
Stillette, description of, **270**
" death by, 270
Strychnine, a cause of ab'n, 47
" to prevent concept'n, 305
Sucking pumps, to induce abortion, 275
Sulphate of zinc, to prevent conception, 306
Sulphuric acid, in abortion, 223
" " to prevent conception, 305
Swelling of cervix, **200**
Syringe, Essex, as a means to prevent conception, 307
Symptoms of induration, 83
" retroversion, 86
" abortion, 122
Syphilis, a cause of abortion, **37**
" sterility, treatment of, 163   **38**
ulcerat'n, a cause of ab'n, 80
" treatment of, 208
constitu'al, cause of ab'n, 81
local, " " **81**

Table showing frequency of abortion, 26
Tabular view of the three great periods of utero-gestation, 309
Tampon, its use in abortion, 224
Tanacetum vulgaris, a **cause of** abortion, 111
Terebinth, a cause of abort'n, 110
Texas, laws of, on " 330
Third stage of pregnancy, abortion at, 247
Third stage of pregnancy, abortion at, sequelæ, 250
Thyroidal causes of abortion, 56
" " treatment of, **179**
Treatment of abortion, 155

Tubercular deposits in placenta, a cause of abortion, 44

Ulceration of the cervix, **a cause** of abortion, 71
Ulceration of the cervix, treatment of, 198
Ulcer, varicose, cause of ab'n, 76
" treatment of, 204
fissural, a cause of ab'n, 77
" treatment of, 205
follicular, a cause **of** ab'n, 79
" treatment of, 206
corroding, a cause of ab'n, 80
calcareous, " " 80
syphilitic, " " 81
" treatment of, 208
Unnecessary abortion, murder, 313
Uterine causes of abortion, 59
" displacements, treatment of, **212**
Utopian theories, 291
Uterine sound, 289
Use of speculum, a cause of abortion, 73
Ustilago madis, a cause of abortion, 112

Vaginitis, a cause of abort'n, 58
" treatment of, 187
" vesicular, " 188
" granular " 188
Vaginal irritation, " 187
" leucorrhœa, " 194
" " a cause of abortion, 67
Vaginismus, a cause of abor'n, 50
" treatment of, 189
Variola, " 164
" a cause of abortion, 38
Variation of period of quickening, 316

Vesical irritation, a cause of abortion, 57
Vesical irritat'n, treatm't of, 186
Vermont, laws of, on abort'n, 329
Virginia, " " 324
Vomiting, a cause of abort'n, 266

Weight of fœtus, at three to seven weeks, 119
Weight of fœtus, at two to five months, 120
Weight of fœtus, at six months to full term, 121
Wisconsin, laws of, on ab'n, 329

Yellow fever, a cause of ab'n, 39

Zymotic diseases, a cause of abortion, 37

www.ingramcontent.com/pod-product-compliance
Lightning Source LLC
Chambersburg PA
CBHW032355230426
43672CB00007B/709